THE WORLD'S
BEST-SELLING
SMARTPHONE

4 4S 5

iPhone®

MADE EASY

KIERAN ALGER AND CHRIS SMITH

FLAME TREE
PUBLISHING

CONTENTS

It's often difficult to know where to start if you have never used an iPhone before. This chapter will introduce what you can accomplish with your new smartphone; it will also furnish you with the knowledge needed to get past the initial set-up process and to get comfortable with the device. You'll become accustomed to using the touchscreen, moving around the iOS software, opening apps and accessing some of the iPhone's basic features.

USING THE PHONE 54

Although the iPhone is a multifunctional device, harnessing the power of several gadgets rolled into one, at its core it is still a communications tool. Chapter two offers a comprehensive guide to making and receiving phone calls, setting up voicemail, importing your contacts, sending and receiving text and picture messages, and even video chatting with your friends and family.

GETTING CONNECTED 92

With the iPhone, the entire world is in your pocket. In this chapter you'll learn how to browse the internet using the Safari app, send and receive email, control your calendar, and post updates to Facebook and Twitter. Chapter three will also explain in foolproof detail how to use the iPhone's powerful location services, which allow you to summon directions instantly, check in at your favourite restaurant, find nearby attractions and explore cities new and old.

Even if you are a smartphone novice, you will have heard of the mobile apps phenomenon pioneered by the iPhone. In this chapter we'll learn how to download and make use of these new and exciting tools from the Apple App Store and we'll discover how they can be used to keep your iPhone feeling fresh and new every single day.

Did you know that your new iPhone is also a camera, camcorder, portable games console, portable music and video player, digital book reader and personal trainer? Chapter five features a comprehensive guide to taking great pictures and video, and sharing them with your friends and family; we'll also explore how to acquire and play your favourite games, music, movies and TV shows, how to read your limitless books and magazines, and how to improve your lifestyle through a host of dedicated apps and services.

ADVANCED iPHONE

Before you know it, you'll be ready to shed that 'iPhone beginner' tag and move on to some of the iPhone's more advanced functionality. Need to restore from a backup? Improve your battery life? Make use of online file storage apps? It's all here. This chapter also features a detailed troubleshooting guide to counter some of the common problems you'll come across when using your iPhone.

INTRODUCTION

As you're reading this book, you've probably just become the lucky owner of a brand new iPhone. If that's the case, congratulations on making the leap into a new world! Now the iPhone is in your hands, you may be wondering, 'Where do I start?'. Don't worry, though, because you've come to the right place. This book offers both a practical and educational guide to quickly mastering this pocket-sized marvel of modern technology. It is suitable to use with any of the models (described on page 9).

Below: The iPhone is many devices rolled into one, plus so much more.

WHAT CAN I DO WITH AN iPHONE?

The iPhone boasts an incredible amount of practical everyday uses for personal and business use. It encompasses a multitude of modern devices but is still small enough to fit in the palm of your hand. It's a phone, a personal organiser, a music and video player, a camera and so much more. This book will help you to master simple tasks, such as making phone calls and sending emails, to more complex tasks, such as taking photos, using apps to edit them and then log on to the internet in order to upload them to your favourite social networks.

A QUICK iPHONE HISTORY LESSON

There have been six versions of the iPhone, with a new model each year bringing additional features and improved software. This book aims to offer a universal guide to users of all iPhone models. If a particular feature only applies to newer or older handsets we'll make that clear. Here's a brief guide to the features and improvements each handset offered down the years.

iPhone Models

Here's a summary of the new features incorporated within each new version of the iPhone to date.

- **iPhone (June 2007):** With the first iPhone, Apple 'reinvented the phone'. They created a new mobile operating system to fit a 3.5-inch, full-touchscreen device. It had a full web browser, an integrated iPod MP3 player, a video player and dedicated applications such as Weather, Calendar and Google Maps.

- **iPhone 3G (June 2008):** The second coming brought a better, 3G internet, and better GPS. The App Store was also introduced, bringing new web-based applications and games.

- **iPhone 3GS (June 2009):** The third iPhone brought a host of speed improvements and included a compass, voice control and a slightly improved camera.

Above: The original iPhone.

- **iPhone 4 (June 2010):** A fresh and critically acclaimed stainless-steel and glass design, a 5-megapixel camera, a front-facing camera for video calls and the new high-resolution Retina display.

- **iPhone 4S (October 2011):** The design remained the same, but the iPhone 4S introduced Siri – a voice controlled personal assistant – and an online back-up solution called iCloud. The camera was boosted to 8 megapixels with full high-definition video recording.

- **iPhone 5 (September 2012):** This is Apple's thinnest and lightest design ever, with a larger 4-inch screen, more power, better graphics, super-fast fourth generation mobile internet (4G LTE), a new charging/syncing connector and the iOS 6 software.

Right: The stainless steel frame of the iPhone 4 acts as its antenna.

The Touchscreen Revolution

The iPhone ushered in the era of touchscreen phones. It dispensed with the physical keyboards – both the full QWERTY keyboards and those awkward numeric keypads – and soon prodding, swiping and pinching a screen accomplished almost everything, save for the power, volume and home buttons.

Apple iOS Software

You'll read a lot about iOS in this book: it's the software that comes pre-loaded on to the iPhone for use straight out of the box. The latest version is iOS 6, which the vast majority of iPhone owners are now using. Apple's improvements to its iPhone handsets each year are always accompanied by a tweaked version of iOS but even if you're using an older phone (3GS and up), you can still upgrade to the new software every time. When certain features are only available in iOS 6, we'll make that clear.

Above: The book will show you how to use Apple's new operating software, iOS, which can be upgraded with all iPhones, 3GS and upwards.

Above: The iPhone 5, released in 2012, is 9mm taller than the iPhone 4, but the same width.

DIVE IN AND DIVE OUT

Rather than reading cover-to-cover, this book has been written in the hope that you will dive in and out when you need a helping hand to understand a particular feature or if you're having trouble overcoming a problem. For example, if you don't know how to download music from iTunes, or email a picture, you can head straight to that page for a detailed explanation. If you're stumped, the easiest thing to do is to look up your topic in the index page.

JARGON BUSTING

While we have made every effort to crush buzzwords and display instructions in the simplest possible terms, sometimes jargon is unavoidable when describing features (iTunes Syncing, Shared Photo Streams, etc.).

Hot Tips

Throughout the book, we have inserted a host of bonus tips to help you get the most out of your iPhone. These simple features can be less obvious or hidden away but can provide the key to unlocking more cool features on your iPhone handset.

Above: The book contains useful screenshots that help you understand what is being explained.

HELP!

We are very confident that the information within this book can help you to become fluent in the language of iPhone in no time at all. However, if you need further advice, help is always close by. The Apple Support website offers hugely detailed archives on how to master each feature and overcome problems..

RESEARCH AND TRAINING

The authors of this book, Kieran Alger and Chris Smith, are both experienced technology journalists and have chronicled the iPhone since it was first launched in 2007. They have used, reviewed and tested every handset to date and both have written countless features and guides on getting the best out of the iPhone, its software and apps.

GETTING STARTED

WHAT IS AN iPHONE?

An iPhone is a mobile phone with software that lets users search the internet, send email, and play music, video and games. You can take photographs, shoot video and find directions. In this section we'll introduce some of the key uses for your new iPhone: there are more than you think!

Above: Once an active SIM card has been installed the iPhone can be used to make and receive calls.

COMMUNICATION

A multi-talented device the iPhone may be, but its primary function is still communication. Here are some of the different ways you can use the device to keep in touch directly with friends and family; all of the following will be explained in more detail throughout this book.

Phone

Once an active SIM Card is placed in the device you'll be able to make and receive calls by hitting the Phone icon on the device's Home screen and using the on-screen dial pad. The iPhone *is* still a phone after all!

Messaging

The SMS (short message service) or 'texting' app is a primary function of the iPhone. Messages can be typed on the touchscreen keypad and, once sent, will appear in a thread, allowing you to keep track of conversations over time. iPhone, iPad, Mac and iPod touch owners can exchange messages with each other for free over Wi-Fi or mobile internet using Apple's own iMessage app.

Email

The iPhone's native email app allows you to send and receive electronic mail directly to your handset; it's easy to configure your Google Mail, Yahoo Mail, Hotmail and more. Microsoft Exchange users can also have their work emails sent straight to the device and we'll explain exactly how to do this in chapter three.

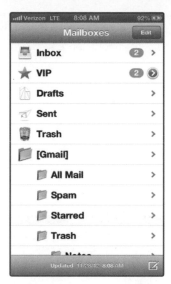

Above: iPhone's built-in mail app can be used to directly recieve and send emails.

Above: The FaceTime app can be used for free video chat with other iPhone, iPad or iPod touch owners.

FaceTime

FaceTime is an Apple-to-Apple free video chat app, which uses the front-facing camera above the iPhone display (iPhone 4, 4S and 5 only). It means you can contact anyone using an iPhone, iPad and iPod touch (plus most newer Mac computers) for free over Wi-Fi.

INTERNET

The iPhone is the entire internet in your pocket and can be accessed through apps or the built-in Safari browser. Whether it's reading online, posting on Facebook or Twitter, watching YouTube videos or shopping, most things that you are used to doing on your computer can be done on the iPhone.

App Store

Each of the icons on your iPhone's Home screen is an app (Mail, Phone, Messages, etc.), but there are millions more of these self-contained

Above: iPhone's Safari browser allows you to access websites on the internet.

applications in the App Store, and they are a great way to keep your phone fresh with exciting new content. We'll discuss apps in chapter four, and at the back of the book there is a list of 100 essential apps to get you started.

Maps

With iOS 6, Apple replaced the popular mobile version of Google Maps with its own Maps app. The latter allows users to search for directions and for nearby places of interest but can also replace your sat nav unit with its voice-guided, turn-by-turn navigation feature.

Above: The iPod MP3 player is incorporated into the iPhone allowing you to download and buy music.

MULTIMEDIA

The iPhone is also a full-on personal media player and games console packed into a pocket-sized device. We'll be going into greater detail on all of these features throughout this guide.

Music

The first iPhone incorporated the revolutionary iPod MP3 player, which allows users to load their digital music collection on to the portable device. You can also buy music from the iTunes Store, or use music streaming apps like iHeartRadio to listen to tunes over the internet.

Video

As with music, you can also transfer your favourite digital movies and TV shows on to the iPhone for small-screen playback. iTunes offers the chance to buy or rent the latest movies and television shows while you're on the move.

Games

The iPhone has become a popular handheld gaming device thanks to addictive games like Angry Birds. Old favourites, such as Scrabble and Monopoly, are available to play too, while

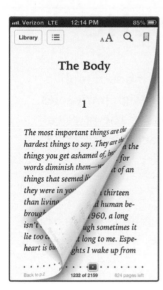

Above: The iBooks app allows you to access a bookstore through which millions of titles can be purchased, dowloaded and read on your iPhone.

versions of console games like FIFA Soccer and Call of Duty can also be downloaded.

Books

Electronic books, or ebooks, are now starting to outsell those made from dead trees, and Apple offers an app and accompanying bookstore (iBooks – *see* page 153) that allows access to 1.5 million books, available to download and be read directly on the iPhone.

CAMERA

The camera on the iPhone 5 is so good that you can probably start leaving your trusty compact camera or camcorder at home on most occasions and not worry about precious memories being tinged by terrible photos.

Hot Tip
Use the Photo Stream portion of iCloud to save your photos online automatically.

Photography

The stills camera has got progressively better since it was an afterthought on the first iPhone (only 2 megapixels!). The iPhone 4S and the iPhone 5 have 8-megapixel cameras, while the iPhone 4 had a respectable 5-megapixel sensor capable of great photos.

Video Camera

The video camera on the iPhone 4S (and iPhone 5) allows users to record full high-definition video at 1080p (a resolution of 1920 x 1080), the same as most video cameras on the market and the same as television shows and movies you see in hi-def.

ANATOMY OF AN iPHONE

The new iPhone 5 arguably represents the pinnacle of modern mobile technology. In this section we'll explain some of the design intricacies, the features within the device itself, and the functions of the physical buttons and switches. But first, let's crack open the packaging.

WHAT'S IN THE BOX?

The first time you see that small box you'd be forgiven for thinking: 'Is this what all the fuss is about?' However, within that minimalist packaging is everything you'll need to harness that new smartphone sitting invitingly within the box.

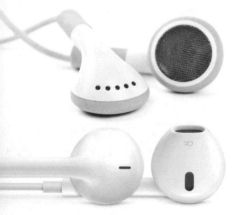

Above: Apple earphones (top) and Apple EarPods (bottom).

Earphones

Apple gives you a free pair of earphones with every iPhone for listening to media and hands-free calling. If you have bought an iPhone 5 you'll get EarPods (designed to fit the ear canal better, and also featuring a microphone and volume controls), whereas earlier versions of the iPhone came bundled with the old style earphones. You can see the difference between EarPods and the headphones offered with previous iPhones to the left.

Charger and Cable

With moderate use, the iPhone's battery should get you through the day, but it will need replenishing overnight. The charging cable slots into the mains adapter while the other end slots into the bottom of the iPhone. The cable can also be plugged into the USB port on your computer in order to charge, sync or transfer content between the latter and your iPhone (see page 21). Apple launched a new Lightning cable with the iPhone 5.

iPHONE EXPLAINED: BUTTONS

Now that the box is empty let's take a look at that beautifully crafted handset. Although most of your activities will be conducted via the domineering touchscreen (opening apps, typing, taking photos, playing music), the iPhone still has a few essential buttons and switches.

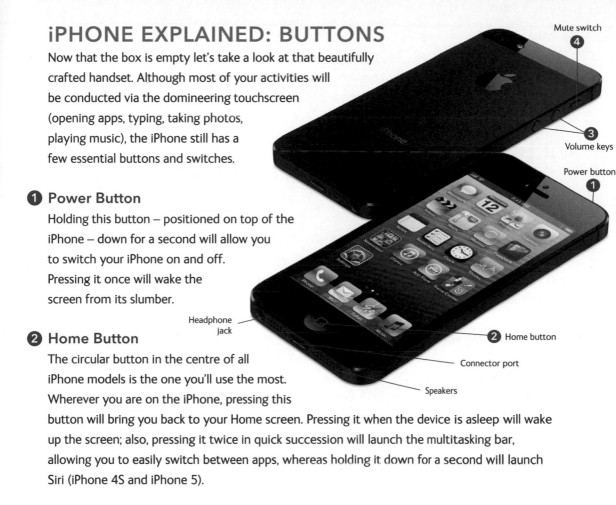

Mute switch

4

3

Volume keys

Power button

1

Headphone jack

2 Home button

Connector port

Speakers

❶ Power Button

Holding this button – positioned on top of the iPhone – down for a second will allow you to switch your iPhone on and off. Pressing it once will wake the screen from its slumber.

❷ Home Button

The circular button in the centre of all iPhone models is the one you'll use the most. Wherever you are on the iPhone, pressing this button will bring you back to your Home screen. Pressing it when the device is asleep will wake up the screen; also, pressing it twice in quick succession will launch the multitasking bar, allowing you to easily switch between apps, whereas holding it down for a second will launch Siri (iPhone 4S and iPhone 5).

❸ Volume Keys

The circular buttons, labelled with + and − icons, are handily placed on the left side of the handset and allow you to adjust the volume of calls and media. In a lot of cases this can also be achieved by using a sliding bar on the touchscreen.

Hot Tip

The volume buttons can also be used to trigger the camera (*see* page 17).

④ Mute Switch

There are few things more embarrassing than your *Star Wars* ringtone blaring in the classroom, cinema or during an important meeting. You can quickly curtail 'The Imperial March' by flicking the Mute switch above the volume keys.

iPHONE EXPLAINED: OTHER PHYSICAL CHARACTERISTICS

① Screen

Up until 2012, all iPhone models had a screen that measured 3.5 inches diagonally. This changed with the launch of the iPhone 5, which brought a new 4-inch design and is therefore longer, but not thicker, than its predecessor. This means widescreen video and room for another row of apps.

② Retina Display

The iPhone 4, 4S and 5 all feature a super high resolution screen that Apple calls the Retina Display; as a result, photos, text and videos appear much clearer to the eye. It basically means that these screens have more pixels than the human eye can distinguish when the phone is held at arm's length.

Headphone Jack

With the iPhone 5, Apple moved the headphone jack to the bottom on the device

Rear camera   FaceTime camera

④   Flash   ③

① Screen

② Retina display

(in all previous versions it was on the top; (*see* image on page 19). You can use this to plug in the bundled-in headphones, your own cans or any speaker that has a 3.5 mm connector cable.

Connector Port

The connector at the bottom of the device allows you to charge your iPhone's battery, and to plug it into your computer to transfer content and synchronize important data (*see* image on the previous page). As explained previously in this chapter, Apple introduced a new connector with the iPhone 5 (*see* image on page 19).

Speakers

There are two speakers on the iPhone 5: one in the earpiece, which allows you to hear calls, and another, more powerful one, on the bottom of the device which plays ringtones, music and other audio (*see* image on page 19).

❸ FaceTime Camera

On the iPhone 4, released in 2010, Apple introduced a front-facing camera, which enabled video calls over its FaceTime app (*see* page 15) but can also be used for stills photography like self-portraits.

❹ Rear Camera

The back of an iPhone is featureless, save for the camera. The lens, which is much smaller than the one in a compact camera, appears in the top left corner of the device. The iPhone 4 also incorporated a flash for the first time, which sits next to the lens.

Hot Tip

There are a host of apps on the App Store that allow the camera flash to be turned into a torch – perfect for finding your keys at night!

Use Protection!

The iPhone isn't a cheap piece of kit to replace. To help keep it pristine, we'd suggest buying a case to safeguard it against drops and a screen protector to guard against scratches and scuffs.

INSIDE YOUR iPHONE

The iPhone may look pretty, but it's on the inside that the magic is created. Each generation of iPhone has become progressively faster, more powerful and more versatile. Below is a list of some of the internal features.

- **Processor**: The iPhone 5 features the Apple A6 processor, which is twice as fast as the A5 processor in the iPhone 4S, which was twice as fast as the A4 processor in the iPhone 4 ... You get the idea.

- **Storage**: The last three iPhone models (4, 4S and 5) have given three storage options: 16GB, 32GB and 64GB. The larger the capacity, the more music, video, apps and photos you can store on the device.

- **Battery**: Unlike almost every other smartphone on the market, the iPhone's battery is not removable.

- **Wi-Fi**: Providing you have the password, this allows you to connect to any wireless network and access internet-based content.

- **Mobile internet**: The iPhone has mobile data connectivity that allows you to access the internet on the go. Your mobile network (O2, Vodafone, etc.) will place a limit on how much data you can use each month.

- **Bluetooth**: This is still one of the best ways of sharing photos or connecting to other devices with your phone. If you have Bluetooth speakers, you can use them to play music wirelessly on your iPhone.

- **GPS**: The chip inside your iPhone allows satellites to pinpoint your location and use mapping services.

GETTING STARTED

Now we're familiar with the iPhone, both inside and out, it's almost time to push that Power button for the first time. In this section we'll get you past all of the tedious pre-use steps and the boring, but necessary, setup screens.

Above: Different iPhone models require different sized SIM cards. You will be provided with the relevant SIM card when you purchase your phone and it is possible to transfer your previous phone number and details across.

THE SIM CARD

If this is your first iPhone, you'll almost certainly need a new SIM Card. The SIM Card is the fingernail-sized piece of plastic embedded with a chip that contains all of your personal information and communicates with your mobile network. The iPhone 5 requires a unique SIM Card called a nano-SIM, which is the smallest yet, whereas the iPhone 4 and 4S use a slightly larger micro-SIM.

Activating the New SIM Card

When you purchase your iPhone in-store or online, you should be provided with a new SIM card. Your network should take care of transferring your current phone number and details over to the new SIM, allowing you to just insert-and-go. In some circumstances it may be necessary to call the network to activate the new SIM. They may ask for the SIM card number and iPhone serial number. It's always better to do this *before* you attempt to activate your iPhone.

Hot Tip

If you insert the SIM and then receive an 'Activation failed' message on screen, you'll know it hasn't been activated yet. Call your network: it'll take minutes to fix.

Above: A pin or SIM tray eject tool can be used to open the SIM tray. After placing the new SIM card face down into the tray, the SIM tray can be pushed back in until firmly closed.

Inserting the SIM Card

As with SIM activation (see page 23), if you have bought or intend to buy the iPhone in-store then the staff there will be more than happy to help you through these steps.

1. Place the SIM tray eject tool (you'll find one in the box or, failing that, you can use a bent paperclip) into the tiny pinhole midway up the right side of the device.

2. Slowly withdraw the pin to release the SIM tray.

3. Remove the SIM Card from its plastic housing and place the SIM Card into the SIM tray so the chip faces down.

4. Carefully replace the SIM tray until it's firmly closed.

POWER UP

Out of the box, the iPhone arrives with a moderate amount of battery charge. Simply hold down the Power Button on top of the phone for a second and you'll see the Apple logo, which stays in place for about 20 seconds.

Above: When you first turn on your new iPhone you will be greeted with the iPhone set-up screen. Slide the arrow across and follow the instructions to set up your phone.

SETTING UP YOUR iPHONE FOR THE FIRST TIME

Your new iPhone only takes a few of minutes to set up. Once the Apple logo disappears, you'll be greeted with the iPhone lock screen. Place your finger at the left edge of the 'slide to set up' bar and drag the arrow to the right. Here's a step-by-step guide to each of the screens you'll encounter, which may vary slightly depending on which iPhone version you are using.

Language and Country

The first two screens ask you to select your language and country of residence. English is selected as default, so if that's your preference, touch the blue arrow in the top right corner of the screen to move forward. On the Country or Region screen, use a finger to scroll down to United Kingdom (or wherever) and touch that country; a tick will appear next to it. Then press the blue Next button.

Above: Scroll down to select your country of residence and then press 'Next' in the top right-hand corner of the screen.

Above: You can select a Wi-Fi network and enter the password. This will then become the default source of internet whenever possible.

Choose a Wi-Fi Network

Next up, the iPhone will ask you to configure a Wi-Fi network to assist with the setup. You can do this through your mobile internet data, but we'd recommend using Wi-Fi if possible.

1. After a few seconds of scanning, the screen will display available Wi-Fi networks and their respective signal strength. Those which have a padlock icon next to the signal indicator mean that you'll need a password for access.

2. Touch the network of your choice. It'll either be named after a place (e.g. Starbucks, Home Network) or retain the codename written on the router (e.g. NETGEAR ZW52).

3. Enter the password, which will also be written on your Wi-Fi router (if you're at work or in a café, ask the IT guys/staff nicely), by typing it in using the on-screen virtual keyboard.

4. Press Join to move on to the next screen. If the password is incorrect, you'll be asked to enter it again.

Connect to iTunes (optional)

One of the newer features in iOS is the ability to set the phone up completely over the air, meaning that you don't have to connect to a computer anymore. However, the 'Connect to a Wi-Fi network' screen still gives you the opportunity to 'Connect to iTunes' and continue the setup that way, as explained below.

1. Plug the provided charging cable into the bottom of the iPhone and plug the other end into the USB port of your desktop/laptop.

2. If you have it installed already, iTunes will launch. If you don't have it or need to update to a newer version then go to www.apple.com/itunes and follow the download and installation instructions.

3. Your iPhone should now show up in navigation bar within iTunes (it's at the top in iTunes 11 and on the left in iTunes 10). Follow the on-screen instructions to complete the setup.

4. You can also back up your iPhone, and transfer music, videos, apps and photos using iTunes, but there'll be more on that later in this chapter.

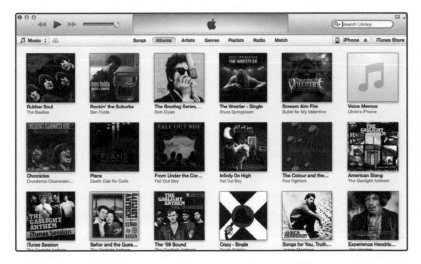

Location Services

Next, you'll be asked whether you'd like to enable Location Services. This is important if you'd like to use the Maps apps to search for directions,

Above: Connect your iPhone to your laptop or desktop computer using the charging cable in order to transfer music you may have stored in iTunes to your new iPhone.

or to use apps that rely on knowing your location. It can also be used to 'geo-tag' photos and social networking posts (e.g. when someone uses the 'Check In' feature on Facebook).

Set up as New Phone

Excuse us for assuming the obvious but the fact that you've bought a book called *iPhone Made Easy* would suggest that you are a newcomer to this technological wonder. So, from the Set Up iPhone screen you should select the Set Up as New iPhone option and press Next.

The other options are Restore from iCloud Backup and Restore from iTunes Backup. These are great options if you've owned an iPhone before or need to restore the device (*see* page 42 for a guide on how to back up your phone, whereas chapter six has information on how to restore your iPhone).

Apple ID

If you want to purchase apps and media, and use things like iCloud, FaceTime and Game Center on your iPhone, you'll need an Apple ID. If you have ever bought anything from the iTunes Store (perhaps for a previously owned iPod?) you will have one.

Above: Set up your Apple ID to start accessing the App Store, iCloud and iTunes.

1. Select Sign In with an Apple ID and enter the email address and password using the on-screen keyboard. Accept the Terms & Conditions on the next screen and move on.

2. If you don't have an Apple ID, select Create a Free Apple ID (*see* chapter five, page 181) to register your details and how you intend to pay for the apps and services you'll buy.

3. If you don't want to set up an Apple ID at this time, you can select Skip This Step and configure it later using the Settings app.

iCloud

iCloud is Apple's free online storage and back-up service. Its most basic function is to ensure that your contacts, documents, photos, mail and calendar are saved on servers in case your device is lost or stolen. Beyond that it's great for remembering the apps, music, books and video you've purchased from iTunes and allowing you to re-download them at your convenience. We'll explain much more about iCloud and its multitude of uses later in the book, but at this point all you need to do is to choose between: Use iCloud or Don't Use iCloud. There is no reason not to use iCloud here.

Above: Follow the onscreen instructions to set up iCloud for free online storage and back-up.

Find My iPhone

Selecting Use Find My iPhone will allow you to track the geographical positioning of the device should someone swipe it or you leave it on the bus. It'll also allow you to lock and wipe the device using iCloud.com, thus safeguarding all of your data. Accept this option: you have nothing to lose – except your iPhone.

Above: If you have an iPhone 4S or 5, you will be asked if you want to enable Siri, which is a voice-controlled personal assistant.

Hot Tip

If Siri has trouble understanding you, it may be your accent. Go to Settings > General > Siri > Language to select English (UK) instead of English (US).

Siri or No Siri?

The next setup screen asks whether you'd like to enable Siri: the voice-controlled personal assistant available on the iPhone 4S and iPhone 5. We'd recommend you give it a try (*see* page 100 for an introductory guide to using Siri).

Diagnostics

The final setup screen (hooray!) asks if you want to automatically send information on how you use the phone to Apple so it can improve its services. Those who worry about privacy should reject this option. There's nothing in it for us, just Apple.

Setup Complete

The final screen thanks you for setting up your iPhone. Selecting Start using iPhone will take you to the Home screen. There's so much more still to set up, such as your email and social networks, but these will be tackled later in the book. For now, let's explore your new iPhone.

iPHONE BASICS

Over the course of the next few pages we'll offer some basic tips on familiarizing yourself with the Home screen, moving around the device, using the touchscreen, typing on the keyboard and more.

THE iPHONE HOME SCREEN

The screenshot below illustrates the iPhone 5's Home screen after you've first set up the device; below are some of the key elements to take note of.

1 iPhone Title Bar
The iPhone title bar features many vital indicators, described here from left to right.

2 Signal strength: The more bars you see the stronger your mobile signal. This affects your ability to make clear voice calls and send texts.

3 Network identity: Here you'll see the name of the network you're registered to (O2, Vodafone, Three, etc.).

4 Internet: If you're connected to Wi-Fi you'll see the fan icon. Once again, the fullness of the fan represents the strength of the signal.
If you're using mobile internet, you'll usually see the letters 3G or LTE. If there's zero connectivity, you'll see a small circle.

Above: Home screen.

5 **Time**: Displays the current time.

6 **Battery**: The battery icon shows, proportionately, how much battery you've used. Once you get to 20 per cent or lower, it will turn red.

Hot Tip

To view battery life as a percentage in the title bar, hit Settings > General > Usage > Battery Usage and toggle the Battery Percentage switch to 'On'.

Other Title Bar Indicators

Here are some further indicators that may occasionally appear within the title bar, depending on what's running on the iPhone.

Bluetooth: This will only appear when using Bluetooth (see page 219). The Bluetooth icon is a **B** infused with an antenna.

1 **Location**: If you're using Maps or your iPhone is scanning for your location, the compass arrow will appear in the title bar.

2 **Play**: If music or video is playing, a play icon will appear.

3 **Clock**: If you've set an alarm, a stopwatch or a timer, a clock icon will be present.

Airplane: If you have enabled Airplane Mode, which disables all mobile and wireless functionality while still allowing you to use other features when flying, a plane icon will appear in place of the mobile signal bars.

Compass arrow   Play   Clock

1   **2**   **3**

Above: Maps screen with compass arrow.

APP ICONS

The majority of your iPhone's Home screen is taken up by rows of apps. There are four rows of four apps on the 3.5-inch iPhones (iPhone, 3G, 3GS, 4, 4S) and five rows of four apps on the iPhone 5.

Above: You can change the default apps that reside in the dock on your homescreen.

> ## Hot Tip
> If there's a red circle with a number next to an app icon, it means you've received a notification. It could be an email, message, missed call or news of an app update.

Dock

At the foot of the Home screen you'll see the dock which maintains the same four icons at all times. The idea is to include the apps that you use most, in order to make them more accessible. The apps that appear by default are Phone, Mail, Safari and Music.

> ## Hot Tip
> In order to change the apps that reside within the dock, hold your finger down on the Home screen on any app on screen. All apps on the screen will start to wiggle. Hold the app that you'd like to replace and drag it out of the dock; then grab the replacement and drag it into the dock.

Above: The more apps you get, the more home screens will open up. You can swipe left and right navigate between them and access the apps.

Multiple Home Screens

The iPhone arrives with two screens of apps but when they're full, more will open up. In order to move between them, swipe a finger left or right across the touchscreen. The white dots beneath the bottom row of apps indicate which screen you are currently on.

Hot Tip

Press the Home button at any time to return to the first screen of apps.

Spotlight Search Screen

If you swipe left from your first screen of apps, it will summon the Spotlight Search screen, featuring the virtual keyboard and a bar that says Search iPhone. If you type 'Dave' it will search your Address Book, Calendar, Email, Messages, Music, Video and other apps on the device for that keyword. Touching one of these results will take you directly to that app, so it's a great short cut for finding precise information on your phone. It also gives you the chance to search the web, which will open Safari and search for the term 'Dave' in Google. The final option is to search Wikipedia, the online encyclopaedia.

Above: You can use the Spotlight Search screen to search your phone and the internet.

BASIC iPHONE NAVIGATION

Now that we're familiar with the iPhone's Home screen(s), it's time to start prodding stuff and learning a few of the basic touchscreen gestures.

Opening Applications

As you've probably already noticed, each of the four-sided icons on your Home screen is an application. Simply touching one of these icons will open that app, full screen, on your iPhone. Try it, you might like it.

Going Home

Whatever you're doing with your iPhone, pressing the physical iPhone button will take you right back to the Home screen.

Summon Siri

You can bring up the iPhone's personal voice assistant by holding down the Home button for a couple of seconds. The phone will make a unique sound to denote the launch and a microphone will appear at the foot of the screen. Press the microphone icon to begin using Siri (see chapter four for more on how to harness Siri).

Above: iOS models feature a multi-tasking bar which makes it quicker and easier to move between apps.

Multitasking

iOS features a neat, time-saving multitasking bar to allow you to move between applications without returning to the Home screen. Double-click the Home button to reveal a row of the apps you currently have open. Use the 'swipe' gestures to find the new app and press it.

Swiping

As we mentioned in the previous few pages, moving your finger across the page to move to the next screen is a key means of navigating your way around the iPhone and will become second nature.

Scrolling

If you're reading a web page, email, a thread of text messages or moving around an app and wish to move up and down the page, use a finger to flick the screen up or down. The faster the flick, the faster the page moves.

Zooming

The iPhone enables many multi-touch gestures, which involve using one or more fingers at the same time. The most popular of these is zooming. You may use this in the camera app to get a closer look at photos or to resize web pages to make it easier to read text.

○ **To zoom in:** Place your thumb and forefinger on the screen and move them both away from each other.

○ **To zoom out:** Place your thumb and forefinger on the screen and move them towards each other in a pinching motion.

Accelerometer

The iPhone has a built-in sensor that knows when you switch the view from portrait to landscape – essentially turning the handset on its side. This is called an accelerometer. It works really well with apps that require typing, as it increases the size of the keyboard, on some websites that have greater readability in landscape mode, and while viewing pictures and videos in the Camera Roll app. Throughout the book, we'll point out times when the accelerometer comes into use.

Above: Videos can be viewed in portrait (above) and landscape (right).

Above: Turning the phone to landscape increases the size of the messaging keyboard.

The iPhone Keyboard

The iPhone's virtual keyboard has a QWERTY layout, just like a PC or laptop. When typing emails, messages, searching the internet or more, you simply touch each letter to enter it. Here are the other basic keyboard buttons.

- **Shift:** Hit the upward arrow once to type a single capital letter. Hit it twice in quick succession and it'll turn blue and turn on the Caps Lock.

- **123:** This button switches the keyboard from letters to numbers and symbols.

- **Microphone:** Dictate the message to your iPhone, rather than typing it.

- **Space:** Just like the space bar on your computer, this allows you to move from one word to the next.

- **Return:** Create a new line on which to type.

- **Delete:** Touching the cross enables you to delete any entered text. Pressing it once will delete a letter, whereas holding it down will delete at a faster rate.

Lock the Screen/Phone

If you're not using the phone, pressing the power button once will turn off the screen and lock the phone; if you're playing music, this will not affect playback.

Waking the iPhone

If you've locked the screen or it has timed out due to inactivity (*see* Auto-Lock on page 40) you can wake it by pressing the power button once and using the Slide to unlock functionality.

If you've chosen a security Passcode (*see* page 39), you'll be asked to enter this. The iPhone will then return to the screen you were viewing when the device was locked.

iPhone Lock Screen

There's plenty you can achieve from the Lock screen.

- When **playing music**, double-click the **Home button** to summon a media control panel to pause, skip or adjust the volume.

- Hold down the **Home button** to summon **Siri**.

- Drag the **camera icon** into the centre of the screen to launch the **Camera app** without unlocking the phone.

- If you receive a **message/email/notification**, simply swipe the notification to go straight to it – no unlocking required.

Above: Media controls, the camera app and Siri can still be used from the iPhone lock screen.

Shut Down

In order to switch off the iPhone, simply hold down the power switch for around 2–3 seconds. The screen will go dark aside from a swipe bar, which requires you to 'slide to power off'. You can hit the cancel bar to return to the Home screen.

BASIC iPHONE SETTINGS

Without getting too deep into the nitty-gritty at this stage, here are a few basic settings that you may wish to adjust while using your iPhone. In order to access them, press the grey Settings icon on the home screen.

Above: It's easy to adjust the brightness of your iPhone screen.

Hot Tip

While optimal screen brightness is preferable, it also has a negative effect on battery life, which can be preserved by turning the brightness down.

BRIGHTNESS

You can change this setting by selecting Settings > Brightness & Wallpaper. Touching the blue brightness bar and dragging it left or right will decrease or increase brightness.

WALLPAPER

From within the Brightness & Wallpaper settings menu you can also alter the appearance of your phone. Below the Brightness meter you'll see Wallpaper: on the left is the Lock screen wallpaper and on the right is the Home screen wallpaper. Clicking either of these will allow you to change them with built-in wallpaper or a photo from your Camera Roll. Once you've chosen the new picture, press Set.

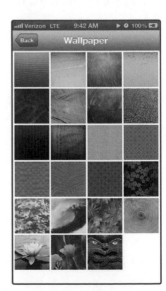

Above: You can select wallpaper options for the Lock screen and Home screen from the built-in wallpaper gallery.

SOUNDS

As well as customizing the appearance, you can also tinker with the sounds emanating from your phone when calls or notifications arrive.

Changing Ringtones and App Alerts

Enter Settings > Sounds to control volume and vibration, and to customize sounds for each volume type. Further down, the Sounds and Vibration Patterns menu allows you to control the precise sound and vibration style for each type of notification. For example, press Text Tone to bring up all of the available options. Selecting one of the options will place a tick next to the name and play a preview. Once you're happy with your choice, you can press the Back button and your selection will be saved.

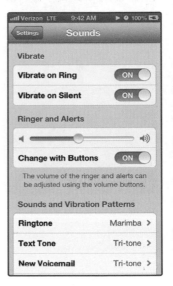

Controlling Ringer Volume

To adjust the volume of your ringtone, simply drag the on-screen slider left or right within the Settings > Sounds menu. It's also easy to control the ringtone volume by hitting the physical volume keys on the side of the device when on the Home screen. Pressing the + or − buttons will bring up an on-screen indicator showing the volume meter increasing or decreasing.

Above: You can select sound and volume options for your incoming texts, calls and other alerts.

SECURITY

Losing a smartphone can be more dangerous than losing your credit cards, bank details and address book in one fell swoop. Here's how to protect your data.

Passcode

Adding a four-digit passcode stops unwanted guests accessing your iPhone beyond the lock screen.

Hot Tip

Don't use numbers like your year of birth, 1234 or 0000 for a Passcode. It's too obvious and the first thing an intruder will attempt.

Above: You can increase the security of your iPhone by choosing a passcode. You will need to enter the passcode each time you unlock your phone.

However, it does mean you need to enter it every time you wish to access your phone from Lock. Enter the Settings app on the Home screen (the grey icon with the cogs) and select General. Scroll down to Passcode Lock and then touch Turn Passcode On. This will ask you to type a four-digit number and then confirm it. If anyone attempts to enter the wrong code 10 times, the phone will be locked.

> # Hot Tip
> Selecting 'Never' from the Auto-Lock settings screen will mean your screen will stay on unless you manually hit the power button. This will drain your battery fast.

Auto-Lock

Within Settings > General > Auto-Lock you can configure the period of inactivity necessary for the phone to lock itself. The default is 1 minute, but you can choose from between Never to 5 minutes.

SYNCING CONTENT ON YOUR iPHONE

There are a number of ways to ensure that the information you carry around with you on your iPhone is up to date. You can sync information using iTunes and iCloud; both have their merits and we'd advise you make use of both.

Syncing iTunes Via USB

The traditional way to sync content on an iPhone is to plug it into your computer via the bundled-in USB cable. Once you plug the device in, the iTunes programme should launch (*see* downloading iTunes on page 180). You'll see your iPhone appear as a button in the navigation bar in iTunes 11 (it appears on the menu on the left in earlier versions). Click the iPhone and select the Sync option at the bottom of the Summary tab.

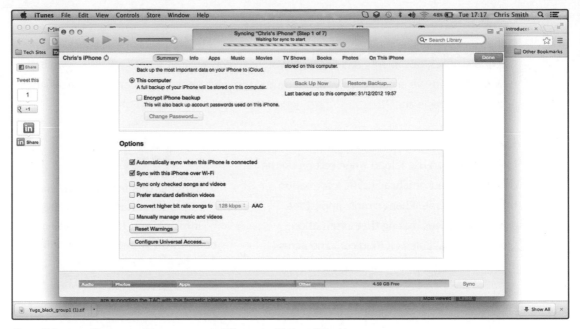

Above: To keep your iPhone up-to-date you can sync it to iTunes using Wi-Fi or a USB cable.

Syncing Via iTunes Over Wi-Fi

If you have iOS 5 (or above) loaded on to your iPhone and you're using iTunes 10.5 or later versions (go to iTunes.com to update to the latest one: iTunes 11), you can sync content over Wi-Fi when both your computer and iPhone are on the same Wi-Fi network.

1. Plug the iPhone into your computer; iTunes will launch. Select your iPhone from the top navigation menu.

2. In the Summary tab, scroll down to the Options section.

3. Tick the box which says 'Sync with this iPhone over Wi-Fi' and hit Sync to save the changes.

4. iTunes syncing will now take place when your phone and computer are on the same network.

What Information is Synced Via iTunes?

iTunes syncs all of the applications, music, TV shows, movies, web bookmarks, books, contacts, calendars, notes, documents and ringtones you have downloaded. iTunes will also download new content you've acquired via your phone and vice versa, ensuring a consistent experience across your phone and PC.

Syncing Your iPhone Via iCloud

If you agreed to use iCloud when setting up the device, it will automatically sync information from some of the iPhone's most important applications, thus making that information seamlessly available at iCloud.com and across any other Apple devices you may have.

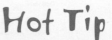

Hot Tip

Syncing this information via iCloud will not back up the information. If you delete a contact on your iPhone, it will be deleted from iCloud too.

Above: You can choose which aspects of your iPhone you wish to sync via iCloud.

What Is Synced Via iCloud?

Enter Settings > iCloud to choose which information you'd like to sync to your iCloud account. Among the options are: Contacts, Calendars, Reminders, Safari (e.g. browsing history and bookmarks), Notes, Photo Stream and more. Turning these on in the iCloud settings will mean any changes you make are saved via iCloud.

BACKING UP YOUR iPHONE

A backup allows you to restore your most important information if you need to perform a full reset or if you buy a new one and wish to pick up where you left off on a new device. You can choose to back up via iTunes *or* iCloud.

Backing Up Via iCloud

Apple gives you 5GB of free storage (you can buy more if you're a heavy user) to store important data in the event you have to perform a full reset. If you choose to enable backups, hit Settings > iCloud > Storage & Backup. Scroll down to iCloud Backup and toggle the switch to 'on'. If enabled, iCloud backups are automatically made when your phone is on a Wi-Fi network, plugged in and locked. In order to make an iCloud Backup manually at any time, you can select 'Back Up Now' from the Storage & Backup screen.

Backing Up Via iTunes

The default setting is to back up via iCloud, but if you prefer to have everything physically backed up on your computer, you can do so via

Above: Backing up using iCloud.

Above: You can back up your iPhone data to your computer using iTunes.

Hot Tip

For everything you need to know about restoring from an iTunes or iCloud backup, *see* chapter six.

iTunes. Plug your phone into your computer and iTunes should load.

1. Select the iPhone in the iTunes menu.

2. From the Backup menu, tick the box that says Back up to this computer.

3. Select Apply to save the changes, and then press Sync to back up the phone. This will switch off the iCloud backup.

What Information Is Backed Up?

Backups over iCloud and iTunes will safeguard all photos on your Camera Roll, all of your account settings (email, Facebook, Twitter, etc.), documents, general phone settings (wallpaper, ringtones, etc.), Contacts, Calendar and more, and make it easy to pick up where you left off if you restore from a backup.

iTunes vs iCloud: Which Should I Use?

As there are two choices for backing up your data, it's sometimes difficult to pick the best one for you. Both options have their advantages; the following guide should help you decide.

○ **iTunes:** If you prefer to have information saved on your computer rather than on the internet, choose iTunes. This is also the better option if you wish to store a larger amount of data than the free 5GB iCloud storage or if you'd rather avoid iCloud completely. Restoring an iPhone via USB connection is also faster than via iCloud.

○ **iCloud:** To go truly wireless, you need to use iCloud. This will enable you to restore your device or set up a new phone with all of your information without connecting to a computer. Also, if you're not backing up that much data, Wi-Fi iCloud backups are quick and simple.

NOTIFICATIONS

The well-connected iPhone user is constantly receiving new information. Emails, messages, social networking updates, upcoming Calendar events, Reminders, app updates, missed phone calls and more.

NOTIFICATION CENTER

The Notification Center, which arrived in 2011 with iOS 5, is one of the iPhone's unsung gems. It doesn't appear until you need it, but when you summon it, it will allow you to catch up on all of the notifications you're yet to deal with. In order to reveal the Notification Center, place your finger on or just above the title bar and drag your finger downward – it should now cover the Home screen.

As you can see from the screenshot, there's a range of information presented (including the Weather app) and everything is actionable.

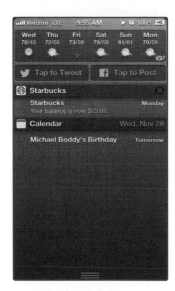

Above: The Notification Center was introduced in Apple iOS 5. It can be accessed by placing your finger on the title bar and dragging it downwards.

This means that touching a notification will take you straight to the email, message, voicemail, etc. You can also post directly to Twitter and Facebook from the Notification Center, but we'll tackle that in chapter three.

Configuring the Notification Center

Enter Settings > Notifications to control which applications appear in the Notification Center.

Hot Tip

In order to remove an item from the Notification Center, hit the small cross on the right side of the message. This won't delete the message, just the notification.

Above: You can select which notification alert you would like for each app.

Above: You can choose how you would like an alert to appear on your screen.

For example, if you don't want Mail notifications to appear then select this from the list and toggle the Notification Center switch from 'on' to 'off'. Likewise, apps that aren't currently configured for Notification Center can be enabled using this method.

Alert Style

Within the Settings > Notifications menu you can also configure the type of alert you'll receive for each app. Select an app from the list and you'll see Alert Style.

- **None:** If you don't want to be altered by this app then select None. This way you'll only see updates from, for example, Facebook, if you manually enter the app.

- **Banners:** A Banner notification appears at the top of the screen and then disappears automatically.

- **Alerts:** Alerts appear in the centre of the screen and require you to act on them by pressing Launch/Reply (to bring up the app) or Close (to deal with it later). These can be quite intrusive for apps like Messages where you may be having a conversation but great for ensuring you see important emails.

- **Sounds:** You can toggle this switch to choose whether the app will make a sound when it receives a new notification.

- **Appear in Lock screen:** Regardless of whether your screen is locked, a notification will wake it up. Selecting this option will ensure that, even if you've not seen it by the time the screen goes to sleep again, it will remain in the Lock screen when you wake it.

ORGANIZING

Have we already mentioned that the iPhone is multi-talented? The device features a host of built-in apps to ensure you never sleep in, never burn dinner and always remember that Eureka idea.

Above: You can add specific events to your iPhone calendar and request multiple reminders.

CALENDAR

The iPhone's built-in Calendar app will sync all of appointments from your email and social networking accounts (e.g. Facebook Events) while new events can be segregated into Work, School and Home sections. Apple's online storage iCloud platform ensures that any new appointments you add using your iPhone will appear across your other Apple devices.

ADDRESS BOOK

Your phone contacts should naturally be imported on to the device when you first insert the SIM card, but it's best to check with your network, as in some cases you may have to back them up first. You can also import contacts from Microsoft Exchange, Facebook, Twitter, Google, iCloud and more. Learn how to manage your contacts in chapter two.

Above: The Contacts app allows you to scroll through your address book and select people to message or call.

CALCULATOR

The iPhone has helped to minimize the number of separate gadgets we need to carry at any one time. The built-in touchscreen calculator, which sits within the Utilities folder on your Home screen, makes it easy to split dinner bills, work out household budgets and even resolve complex formulas.

Above: Always handy for your (or your child's!) homework, the iPhone's scientific calculator is a useful tool.

CLOCK

As well as the clock that sits within the centre of your title bar, there is a standalone Clock app on your Home screen, which includes a World Clock, Alarm, Stopwatch and Timer.

Hot Tip

Turn the iPhone on its side to switch to the scientific calculator.

Above: World Clock lets you view the time across multiple locations.

World Clock

If you have family abroad, or want to work out what time the Super Bowl starts, the World Clock is a handy feature. By default, the iPhone features Cupertino (Apple's headquarters in California), New York and London but you can add other locations to the list.

1. Select Clock > World Clock and then press the + icon. Click in the search box and begin typing a city of your choice, such as Paris.

2. To delete a clock from the list, hit Edit in the top left corner and then touch the red circle and hit Done.

3. To rearrange the order of the clocks, hit Edit and drag the bars next to the clock up or down.

Alarm Clock

Mobile phones have been consigning the trusty alarm clock to the scrapheap for years and the iPhone continues that tradition.

1. Select the Clock app and hit the Alarm icon at the bottom of the screen.

2. Press the + icon in the top right to add a new alarm.

3. At the foot of the screen, use the hour, minute and AM/PM scrollable wheels to choose the time. Press the Repeat option to select which days you'd like the alarm to go off.

4. Select Sound to choose an alarm (choose from default ringtones, a song from your music library or select 'Buy More Tones' to head to the iTunes store).

Above: Alarm Clock allows you to save multiple alarms with your choice of repeat, sound and snooze options.

5. Toggle Snooze on or off to give you the option of an extra 10 minutes.

6. Press Label to give the alarm a name, i.e. work, feed dog, etc.

7. When you are satisfied, click Save to return to the Alarm screen. From there you can toggle the alarm on and off.

8. Selecting the Edit button in the top right corner enables you to delete the alarm completely or change the settings discussed above.

Stopwatch

We remember a time when the stopwatch was a feature to get excited about on a digital watch. Now it's just an afterthought on a smartphone, but still as useful.

1. Hit Clock > Stopwatch and then the green Start button to set the digital display running.

2. Hit the same button to stop or, if you're running around a track, hit Lap to keep it running but record that time. Lap times will appear in the list below.

3. Hit Reset to return the stopwatch to zero.

Below: The timer screen gives you access to sound options as well as displaying cancel and pause buttons.

Timer

Perfect if you've got a joint of beef in the oven or need to set intervals between your next antibiotic, the Timer is the final option within the clock app.

1. Hit Clock > Timer and use the vertical scroll wheels to set how many hours and how many minutes you'd like to count down from.

2. Choose an alert to sound when the timer runs out by selecting 'When Timer Ends' and choosing from available ringtones.

3. Press Start to begin the timer.

4. A new screen will launch, informing you of how long is left and also enabling you to pause or cancel the timer.

> .ıll Verizon LTE   10:02 AM   ▶ ◉ 100% ▣
>
> # 16:54
>
> **When Timer Ends**   Marimba >
>
> Cancel   Pause
>
> World Clock   Alarm   Stopwatch   Timer

Hot Tip

You can exit the Clock app and the timer, stopwatch and alarm will continue to run in the background.

NOTES, REMINDERS AND VOICE MEMOS

The iPhone features a number of ways for you to preserve information you're liable to forget. In some cases, it will even send you endless reminders. Here are three apps that specialize in remembering, so you don't have to.

Notes

The pretty, built-in Notes app is given its own icon on the Home screen and, once opened, takes the form of a classic yellow notepad. All of your previous notes will be listed, accessible and editable with one touch. You can also hit the on-screen + button to start a new note (e.g. shopping list, fancy dress ideas, lecture notes, etc.) and use the keyboard to begin typing away. All notes will be automatically stored without having to save them.

Above: Using the keyboard, the Notes app allows you to automatically store thoughts, ideas and lists.

> ## Hot Tip
>
> In order to sync all of your notes to iCloud, hit Settings > iCloud and toggle the Notes switch to on. You can also change the font by entering Settings > Notes.

Reminders

This excellent to-do list application allows you to check off items and receive alerts for things you haven't done yet. Here is how it works.

1. Launch Reminders and you'll see a notebook with today's date.

2. Click + in the top right corner. This will summon the keyboard, allowing you to begin typing a to-do item (e.g. pick up kids).

3. If this is the only item then press Done but if there's more, simply hit return on the keyboard to add a new one.

4. When you've completed the task, return to the app and tick the box next to the item. This will move the task to the Completed screen which can be viewed by clicking the list icon in the top left corner of the app.

5. In order to set reminders for future dates, touch the corresponding day at the foot of the screen. You can scroll to the right to access dates beyond the next 7 days.

6. In order to set the reminder, select the item and then Remind Me On A Day to bring up a vertical scroll wheel for days, hours, minutes and AM/PM; once finished, select Done.

Above: The Reminders app is an excellent way to organize and keep track of tasks.

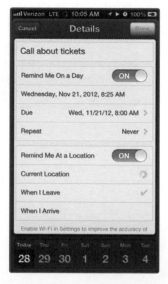

Geo-reminders

Reminders are really cool – you can even select Remind Me At a Location. For example, if you always forget to buy toothpaste, you can ask the iPhone to send you a notification when your GPS sensor detects that you're near the supermarket on a map. In order to achieve this, set the Reminder as explained above and then select the particular item to bring up the Details screen. Select 'Remind Me At a Location' to enter an address or choose a person from your contacts list (hit Current Location and the address and contact screen follows).

Left: You can choose to recieve reminders when at a particular location.

Voice Memos

The Voice Memos app allows you to record conversations, random great ideas, interviews, humorous anecdotes, messages for friends and family, and more.

1. Launch the Voice Memos app from within the Utilities folder on the Home screen.

2. Hit the red record button to begin recording. A red 'Recording' bar will appear in the Title Bar indicating the ongoing length of the recording.

3. The VU indicator will monitor the volume of the sound. It's best to keep it out of the red zone to ensure that the audio retains clarity.

Above: The VU indicator on the Voice Memos app allows you to keep track of recording volume.

4. If you wish to pause your recording at any point, hit the pause icon on the left (and again to restart). Once you've finished recording, press the stop button to the right.

5. All recordings can be accessed using the 'list' button in the bottom right corner. Playback can be controlled from this screen.

Hot Tip

In order to share a Voice Memo via message or email, enter the list, hit the blue share icon and then hit Email or Message to launch the file within those apps.

USING THE PHONE

MAKING CALLS

Despite being a multi-purpose tool, the iPhone is still a phone, used for making phone calls. Basic calling is simple: tap a name in your Contacts, tell Siri to 'call Chris' or touch a name to return a recent call – but there's much more. The latest models support Wi-Fi video and conference calls, and this section will show you how.

CONTACTS

Whether you are choosing who to share photos with or when to accept calls from colleagues, setting up your contacts in a smart way is crucial for making your iPhone work harder for you.

There are numerous ways to add contacts: from importing from a SIM Card to syncing using iTunes. Once imported (see below), you'll find phone numbers, email addresses, Twitter and Facebook info for everyone in your address book. Tap the Phone app, hit the Contacts icon and from there you can apply some simple iOS 6 magic to really make your phonebook work for you.

Importing Contacts

How you add your old contacts on to your new iPhone will depend on where they are stored. The most common ways to achieve this are the following.

○ **From a SIM Card:** Insert the SIM Card containing all your contacts into your new iPhone (but remember that the iPhone 5 requires a smaller nano-SIM so if yours are

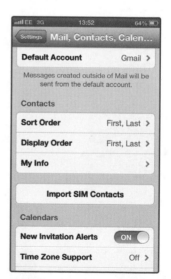

Above: You can import contacts from a SIM card. This is the best way to transfer such information across from your old phone to your new iPhone.

stored on a regular SIM Card you'll need to choose an alternative option). Go to Settings > Mail, Contacts, Calendars > Import SIM Contacts.

Above: You can transfer your contacts from a Google Android phone by saving them to Gmail and then importing them.

- **Switching from Google Android?** You can save your contacts in Gmail and set up your Gmail account on your iPhone 5 via Settings > Mail, Contacts, Calendars. You'll then be given the option to import your contacts.

- **From iTunes:** Copying contacts from an old iPhone? Simply make an iTunes backup and then copy it to your new handset. Connect your old phone to iTunes via USB cable then right-click on your device in iTunes and select Backup. After that's complete, connect your new iPhone via USB, right-click on it in iTunes and select Restore from Backup.

- **From iCloud:** If your old iPhone was running iOS 5 or iOS 6 you can sync your contacts using iCloud. On the old device, go to Settings > iCloud > Storage & Backup and make sure that iCloud is switched on. Then do the same on your new device.

- **From a Microsoft Exchange Global Address List:** Go to Settings > Mail, Contacts, Calendars, select your Exchange account and turn on Contacts.

Hot Tip
Want to see where a contact lives on a map? Open up their contact card and click on the address. This will launch their location in Apple Maps.

Importing Contacts from Facebook

iPhone users who want to connect their Facebook contacts with their iPhone address book now have a superfast way to do just that, provided they're running iOS 6.

Before you do this, though, it's worth deciding whether having all your Facebook friends in your Contacts is desirable. In most cases you'll need to link Facebook entries manually with the contacts already in your iPhone which, depending on how active you are, could be hundreds and clutter up your Contacts.

- **Activate your Facebook account on your iPhone**: Open Settings, select Facebook and enter your Facebook account details.

- **After connecting with Facebook**: Click Update Contacts and this will integrate your Facebook friends' contact details, pictures, email addresses and phone numbers where available. This will also create a Group in your Contacts called Facebook.

Above: By activating your Facebook account on your phone you can integrate 'friends' into your contact list.

Importing Contacts from Twitter

As is the case with Facebook, it's also possible to add your Twitter contacts into your Contacts book. Once again, be aware that this could mean hundreds of people, depending on how active you are, and you'll need to merge Twitter entries with people already in your iPhone manually.

- **Activate your Twitter account on your iPhone**: Open Settings, select Twitter and enter your Twitter account details.

- **After connecting with Twitter**: Click Update Contacts and this will place the Twitter contacts into your Contacts book. You'll usually find them under their full name (or what they told Twitter their full name was).

Merging Duplicate Contacts

Merge duplicate Twitter and Facebook contacts by going into Contacts and selecting one of the entries. Click Edit in the top right corner of the page for that entry and then scroll down to the bottom to find the Link Contact button. Tap this and select the contact you wish to link.

Adding Contacts Manually

There are two main methods for adding new people into your Contacts book.

○ **Brand new contact:** From within Contacts, tap the + button. This will bring up a form for you to fill in with the person's details, such as First Name, Last Name, Phone Number, Email and Address. From here you can also assign things such as ringtone, text tone and photo or apply social networking preferences.

Above: You can add and edit contacts manually, and can choose to include extra details such as an address.

○ **Previous caller:** Open your Recent Call list and scroll to the number you wish to add. Tap on the blue arrow next to it and then hit 'Create New Contact'. You can then enter the relevant information.

> # Hot Tip
> You can manually add a Twitter handle (e.g. @JoeBloggs) to an existing contact by selecting their card in Contacts, scrolling down to the Twitter field and typing in their Twitter username.

SYNCING CONTACTS

Once you have imported your contacts on to your iPhone, the next step is to ensure they're being kept safe and constantly up to date. Fortunately, it's now easier than ever to back up and synchronize your vital information, and to make it readily available to access wherever you are and across all your devices – computer, iPad, iPod or other iPhone.

For those desperate to go wireless, Apple's new iCloud feature (iOS 5 and iOS 6 only) now also lets you sync your iPhone contacts and data over the air, without having to plug the device into a computer, but we'll also explain the traditional method using iTunes and cables.

Choose Your Sync Weapon

Both iCloud and iTunes sync have their pros and cons but it's important to choose one as your standard syncing option. If you alternate, or even do both simultaneously, you may end up with a phonebook full of duplicate data and cleaning that up won't be fun.

Syncing Contacts Using iCloud

Apple's iCloud web-based service now offers computer-free updates for your contacts and data but you'll need to be connected to Wi-Fi before you follow these steps.

1. Tap Settings on your Home screen and choose iCloud.

2. Tap Account and, if you haven't already done so while setting up, add in your Apple ID and password.

3. From here you can choose which of the services you'd like to sync using the On/Off switches. In this instance, you just need to ensure you've turned on Contacts.

Step 2: Whilst connected to Wi-Fi you can sync your contacts using iCloud by selecting 'iCloud' from the 'Settings' menu.

Syncing Contacts Using iTunes

If you prefer to manage you information on a larger screen and close to a power source, then connecting to iTunes via a USB cable is your best bet – here's how.

1. Connect the iPhone to your computer and iTunes should launch automatically. If it doesn't, you'll need to open it manually by clicking on the icon on your computer desktop.

2. Once iTunes is open, select your iPhone from the source list on the left-hand side of the iTunes window.

3. Click the Info tab and scroll to the section Sync Address Book Contacts.

Step 4: By connecting to iTunes and opening the 'Info' tab you can choose to sync all your contacts, or select specific groups.

4. Here you can choose to synchronize all your contacts or identify selected Groups you wish to sync by checking the boxes next to them.

5. If you use Google Contacts, tick the Sync Google Contacts box and click Configure to add your Google account details.

Hot Tip

If you want iTunes to launch and sync your iPhone automatically whenever you connect it to your computer, select the Summary tab and tick the box in the Options section that says 'Open iTunes when this iPhone is connected'.

Searching Contacts

Using the iPhone's tactile touchscreen to scroll rapidly up and down through your Contacts list is probably the most intuitive way to find that person you want to call or email but there are other shortcuts.

- **Jump to contacts by letter**: Use the alphabet on the right to quickly view all the names on your list, starting with 'A'.

- **Simple search**: Start typing the name of the contact you're hunting for into the search bar at the top of the screen. Your contacts will automatically filter to display matches.

Organizing Your Contacts Using Groups

In iOS 6, groups of contacts are created and managed on your PC or Mac rather than directly on your device but once these have been set up and synced to the latter, you can use them to filter what displays in your overall contacts listing – and here's how.

Above: You can search your contacts by typing the name you are looking for into the search bar at the top of the screen. This will quickly filter through your contact list.

1. Enter Contacts and select Groups. At first you're likely to see all Groups listed with a check mark.

2. Press All Contacts to clear these check marks.

3. Next, touch only the Group(s) that you want to display when you open your Contacts. You can select as many groups as you like.

4. Tap Done and all the contacts within the Groups you selected with check marks will now be visible in your Contacts.

Editing and Deleting Contacts

In order to change contact information or delete someone who is now out of your good books, go to Contacts and select the entry you want to amend or remove. Once the contact is open, click Edit and you can freely add, amend and delete details using the green + buttons or the red circles. On the other hand, if you want to delete the contact entirely then scroll to the bottom and hit Delete Contact.

Sharing Contacts

'Can you send me their number?' is a question that gets most people scrabbling through their phonebook, before reading out the answer while someone else taps it into their phone; however, there is a faster way to share contacts. Simply tap the contact you wish to send, tap Share Contact and you can send the info by email or text message.

Above: Once you have set up a 'My Info' Card, any contacts included will be marked in your contact list as with 'Dave Therave' in the above screenshot.

Setting Up Your 'My Info' Card

The information on your 'My Info' card gets shared whenever you send someone your contact details. In order to edit this information, go to Settings > Mail, Contacts, Calendars and hit My Info. Then select your Contact Card. From here you can add, edit and delete information and also adjust how Siri and other apps use your info.

Hot Tip

By default the iPhone sorts contacts by last name but often you only have a first name. To order your contacts by first name go into Settings > Mail, Contacts, Calendar and toggle the Sort Order and Display Order so they say 'First, Last'.

FAVOURITES

In iPhone speak, your Favourites are the numbers you call the most. Just like speed dial on a normal phone, you can store all the vital numbers you'll be dialling regularly for easy access and fewer taps, thus saving time when you want to call.

Calling a Favourite

It couldn't be simpler to phone one of your chosen people: just hit the Phone app icon and then select Favourites from the bottom left corner of the screen. Pick the person you want to call, tap their name and you're dialling.

Adding Favourites

It'd be amazing if the iPhone could automatically fill up your Favourites group based on the number of times you've called people in your phonebook but, currently, anointing a Favourite is still your choice ... Plus, it might be embarrassing to have that takeaway listed at No. 1. With that in mind, there are two methods for adding a new Favourite.

Above: Contact card with 'Add to Favourites' button displayed.

- **When you're viewing a contact**: An Add to Favourites button will appear on the screen. Tapping this brings up all of the phone numbers you have stored for that person. You can then select which number to make your Favourite for that contact.

- **From within your Favourites**: Tap the + symbol in the top right corner. This brings up your Contacts list where you can choose your new best buddy by following the steps above.

Adding More Than One Favourite for a Single Contact

You're clearly going to end up with three different numbers for many people and the good news is that you can also add multiple Favourites for a single contact. To create different Favourites for your partner's office landline, work mobile and personal phone, just repeat the steps above.

Above: Selecting the red circle icon on the 'Favourites' screen will allow you to delete a contact from your favourites list.

Editing Favourites

Managing your existing Favourites is simple. In order to access the full list, simply tap on Contacts from the Home screen and then hit the Favourites button which appears along the bottom of the screen. From there you will be able to carry out the following actions.

○ **Delete a Favourite**: Hit the red circle icon and then tap the Delete button that appears next to your soon-to-be-former favourite. This won't delete them from your Contacts – only from the Favourites.

○ **Re-order your Favourites list**: Just press and hold the icon with the three horizontal lines that appears next to the contact's name. While still holding, drag and drop it into the new position; hit Done when you're finished tweaking the list.

Hot Tip

Only want to receive calls from the special people on your Favourites list after 9pm at night or over the weekend? Select Favourites in the Do Not Disturb settings to filter out all other callers.

DIALLING

Time to make a call! Start by hitting the Phone icon on the Home screen and this will bring up the familiar number keypad, plus four other options along the bottom: Favourites, Recents, Contacts and Voicemail. You can then choose how you want dial.

○ **If you know the number**: Use the keypad to enter the digits manually and then hit the call button.

Above: To call a number if it's not stored in your Contacts, simply dial it by using the keypad, then press 'Call'.

○ **Dialling a recently dialled number or missed call:** Tapping the Recents icon opens your call log, showing all the incoming and outgoing calls to your phone. You can filter these to show All, Missed or Completed calls. From here you can hit the blue + button to see more details of who called or tap the name to return the call.

○ **Dialling using Favourites:** Press the Favourites icon to call up all your personal VIPs and touch anywhere on their name to start dialling.

○ **Calling from Contacts:** Select Contacts from the bottom of the screen. Scan your address book by scrolling, searching or using the A–Z strip and then tap on the contact you wish to call to bring up their Contact card. From here you can choose which number to dial if they have multiple entries (e.g. work, home and mobile).

Above: Selecting the 'Recents' icon on the 'Contacts' screen displays all your incoming and outgoing calls.

Voice Dialling

Controlling your devices using only your silky voice is an emerging trend. Until now it's been fairly unreliable but advances in voice recognition technology mean that smartphones and computers can now decipher us better than ever. Apple's take on voice control – the personal assistant Siri – was developed for iOS 5 and iOS 6, and is leading the way.

○ **Standard Voice Dialling (all versions):** Press and hold the Home button until the Voice Control screen appears and you hear a beep. Say 'Call' or 'Dial' and then say the

Above: Voice control screen button is displayed after pressing and holding the home button.

Above: Call screen with the speakerphone option selected.

name or number you wish to call. Remember to speak clearly and naturally – no need to impersonate a robot or a Brit buying a coffee in Paris. Use full names and add 'at home', 'at work' or 'mobile' where necessary.

○ **Voice Dialling using Siri (iPhone 4S or later models):** If you have Siri activated (Go to Settings > General > Siri), just press and hold the Home button until Siri appears and then say 'Call Bob'. Siri will launch the Phone app and dial for you.

> **Hot Tip**
>
> Prevent voice dialling when the iPhone is locked. Go to Settings > General > Passcode Lock and then turn off Voice Dial. After this you must first unlock the iPhone to use voice dialling.

HANDS-FREE

In the previous section we showed you how to use voice dialling to make a call. If you want to make the entire conversation a hands-free affair so you can apply your opposable thumbs to more important things, such as holding the steering wheel or taking dinner out of the oven before it burns, here's how you can do that with your iPhone.

○ **Put your phone on speakerphone:** From the Call screen, you'll see a Speaker icon. Tap this at any time during the call and the audio will playback through the iPhone's built-in speaker. Tap the icon again to turn off speakerphone and return to your phone's normal speaker.

○ **Use a wired headset**: You can use the Apple Stereo Headset that comes boxed with your phone or you can buy one from another manufacturer. There are hundreds to choose from but make sure you get one with a built-in microphone and a centre button you can press to answer calls.

○ **Use a wireless headset**: You can skip the cables entirely and pair your phone with a wireless Bluetooth headset (*see* page 244). After you've paired it once, any time you're in range of your iPhone (and your iPhone's Bluetooth is switched on) your headset will automatically connect, leaving you free to answer and make calls without ever touching your phone.

○ **Use Bluetooth to pair with your car's sound system:** Most new cars, built in the last five years, come with a Bluetooth-enabled media system. The principle for connecting your iPhone to this is exactly the same as a Bluetooth headset. Make sure your iPhone Bluetooth is switched on and then follow the pairing instructions on your car media system. Once paired, many car systems offer enhanced controls such as phonebook syncing and calls controlled via buttons on the steering wheel.

Above: The bluetooth settings screen can be used to pair other Bluetooth devices with your iPhone.

How to Pair a Bluetooth Device With Your iPhone

1.   Make the device you're connecting with your iPhone discoverable.

2.   On your iPhone, go to Settings > Bluetooth and switch on Bluetooth.

3.   The iPhone will scan for available devices with which to pair and you should see your device listed. The name displayed will depend on what the manufacturer allocated but hopefully it should be obvious. Select your device and, if prompted, enter the pairing code allocated by the manufacturer (see the product manual).

Hands-free Siri (iPhone 4S and later)

In addition to making hands-free calls with the standard iPhone voice control, it's also possible to use Apple's new voice-controlled personal assistant Siri to start and end calls, without having to touch your iPhone.

Hot Tip

If you're prompted for a code while attempting to pair Bluetooth devices, try entering '0000'. This is very often the default code used by device manufacturers. Specific codes are reserved for products where added security is required.

You can also write and send messages, schedule meetings, get directions, set reminders and search the web, simply by talking. Siri works with the headset that came with your iPhone or you can buy your own compatible wired or Bluetooth headset.

Hot Tip

The more Siri knows the better. Create Contact cards for key family members such as Mum or Sister. Add nicknames, addresses, and email addresses to help Siri respond better to your requests.

○ **To start talking to Siri using a headset**: Press and hold the centre button on your iPhone (or the call button on a Bluetooth headset). Siri will come to life and ask what you'd like to do. You can then fire off instructions, such as 'Call Dave'.

○ **To continue a conversation with Siri**: Once you've got Siri's attention, press and hold the button each time you want to talk. When using a headset, Siri will respond to you via your earphones.

○ **Sending hands-free messages**: You can dictate text messages and emails via your headset mic to Siri. She'll read back your beautiful prose before sending, in order to give you a chance to ensure it's all correct.

Above: The screen above will be displayed when Siri is on and working, you can then issue requests such as asking it to send messages.

CONFERENCE CALLS

A conference call allows groups of people to connect on a single phone call. They are mainly used for business meetings but they also provide a great new way for families to catch up over long distances. Provided your mobile phone network allows it, you can create iPhone conference calls with up to five other people.

You can make international conference calls but these often prove expensive. It's advisable to check your data plan before calling an auntie in Sydney and a brother in New York for that weekly catch up. Alternatively, for international calls, you can use services such as Skype, FaceTime or Google Hangouts over Wi-Fi. This not only saves money but also offers the added bonus of video calling.

Managing a Conference Call

Conference calls are often pre-arranged with a designated list of attendees and one person taking responsibility for 'chairing' the conference. However, the iPhone makes it very easy to bring people into existing calls, so you can also create a conference call at those unexpected moments when two family members call you at the same time. Whether you're the person inviting others on to the call or you want to connect people dialling in, here's how you turn a one-to-one call into a conference.

Above: Selecting 'Add Call' allows you to merge others into an existing call.

○ **Make your first call**: For a pre-arranged conference call, you'll need to dial one of the people joining the call first in the usual way (*see* page 66 for help with dialling).

○ **To add a second person**: While on your first call, tap Add Call and then make another call. Next, select Merge Calls. Repeat as necessary to add up to five people to the discussion.

○ **To remove an attendee from the call**: People can obviously leave of their own free will but should you need to eject someone, select Conference, tap next to a person and then press End Call.

Above: When you recieve an incoming call the call screen will display the option to answer or decline the call.

○ **To chat privately with one person during the call**: To have a bit of secret sideline whispering with someone on the call, just tap Conference and then press Private next to the person you want to speak to. Hit Merge Calls when you're ready to resume the conference.

Hot Tip

Make friends with the mute button on your iPhone and use it when you're not speaking so you'll be able to cut out background noise. It also leaves you free to sneeze, cough, heavy breathe or have a conversation in the real world without anyone knowing.

○ **Adding an incoming caller**: If an incoming caller wishes to join the call, simply tap Hold Call + Answer, then tap Merge Calls and this person will be added to your conference call.

RECEIVING CALLS

Taking incoming calls on the iPhone can be as easy as a single tap but with a little customization you can tailor your phone to respond to calls in a way that suits you best, based on who's calling, where you are or what you're doing at the time.

ACCEPTING CALLS

Life is full of decisions and each time your iPhone lights up, rings or buzzes (*see page 46* to learn how to set your alerts) with an incoming call, you've unknowingly been handed some more to make. Ultimately, what you do next rests entirely on whether you're in the mood for talking, you're too busy for chit-chat or you just want to be left alone. Whatever your state of mind, here are all the tools you need to deliver an appropriate response.

○ **Answering a call:** If your phone is already unlocked, just tap the green icon to answer. If the phone is locked, drag the slider and then tap to answer.

REJECTING CALLS

We're here to help not to judge, so whatever your reason is for rejecting a caller, here are the most straightforward ways to decline an inbound call.

Left: The incoming call screen can be used to answer or decline a call, but there are also other ways to reject a call, such as sending it to voicemail, silencing it or replying with a text message.

○ **Decline a call and send it to voicemail**: Tap Decline, press the Sleep/Wake button twice quickly or, if you're using a headset, press and hold the centre button for a couple of seconds until you hear a double beep.

○ **Silence a call**: To ignore a call without rejecting it, press the Sleep/Wake button or either volume button. You can still answer the call after silencing it, as long as you catch it before it goes to voicemail.

○ **Reply to an incoming call with a text message**: Swipe up on the phone icon, tap Reply with Message and then choose a reply or tap Custom to send a carefully crafted brushoff of your own (to create your own default replies go to Settings > Phone > Reply with Message and replace any of the default messages).

Hot Tip

The iPhone doesn't support call blocking.
Silence persistent rogue callers (telesales people,
that's you) by adding them as a contact, then
download and assign them a silent ringtone.
Next time they call, your display will light
up but that's all.

○ **Remind yourself to return an incoming call**:
Remind Me Later is a handy feature on the iPhone that lets you reject a call but gives you a nudge in the ribs to call back when it's more convenient. When a call comes in, swipe up on the phone icon, tap Remind Me Later and choose when you want to be reminded.

Above: The 'Reply with Message' and 'Remind me Later' options appear on the incoming call screen if you swipe upwards on the phone icon.

Do Not Disturb Mode (iOS 6 Only)

Do Not Disturb is a new feature of iOS 6; it's designed to prevent irritating interruptions while you're in meetings, in church or just trying to sleep. Of course, cutting yourself off from the world entirely would mean emergency calls or special loved ones wouldn't get through, but fortunately Apple has also thought of that. Do Not Disturb can be customized to enable exceptions – here's how you activate and fine-tune your built-in call buffer.

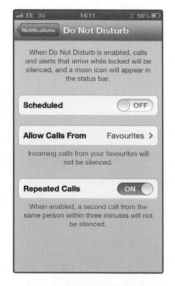

Above: The Do Not Disturb settings screen allows you to customize which calls you receive.

- **To switch on Do Not Disturb**: Go into Settings > Notifications > Do Not Disturb and flick the button next to Schedule to on.

- **To schedule when Do Not Disturb operates**: Tap next to the From and To sections to assign times when you want Do Not Disturb to be switched on.

- **To set exceptions**: Press the Allow Calls From button and then choose from the following options to dictate who gets through your protective call shield: Favourites, Everyone, No One, All Contacts or a specific Group you've set up in your Contacts.

- **Repeated calls**: If you feel like making an exception that allows persistent callers through, you can fire up the Repeated Calls option. If this is switched on, anyone who calls you twice within three minutes will be able to disturb your slumber/ meeting/prayers.

Juggling Calls

If you're busy talking on the phone when you get a second incoming call, the iPhone lets you juggle calls, so you can quickly let the caller on line two know that you'll be with them shortly without hanging up on the person on line one. Equally, you can end your first conversation to take the second call or just ignore the new caller and carry on as you were.

○ **Put your current call on hold:**
Touch and hold the Mute button.

○ **Put your current call on hold while answering a new incoming call:** Tap the Hold Call + Answer button that appears when the second call comes in.

○ **Ignore the incoming call and send to voicemail:** Tap Ignore.

○ **End the first call and answer the new one:** Press End Call + Answer.

○ **On a FaceTime video call:** You can either end the video call and answer the incoming call, or decline the incoming call.

○ **Switch between calls but keep both alive:** Tap Swap and the active call is put on hold.

Above: If you recieve an incoming call while already on the phone you will be presented with the above options.

Hot Tip

Be aware: any Clock app alarms will still sound even when Do Not Disturb is enabled. Also, notifications and phone calls will come through normally if you're using your device (i.e. if the screen is on).

RINGTONES

The iPhone comes with its own selection of chirpy ringtones but the internet is full of thousands more and you can even make your own. Before we continue, though, let's be clear about one thing: it's your duty to friends, family and the world to choose wisely. Put it another way: get it wrong and it's social suicide.

○ **Choosing a ringtone**: Go into Settings and choose Sounds > Ringtone. A scrollable list of available tones will appear. Tap any of the choices for a preview; a tick will display next to the most recent tone you've listened to and the item will appear at the top of the list as your chosen incoming call alert.

○ **Buying and downloading new ringtones**: You can buy new ringtones wirelessly from your phone via iTunes. From your Home screen, select iTunes > Ringtones; here you can trawl tones by genre and popularity. To purchase, click the Buy button and the ringtone will appear in your available list.

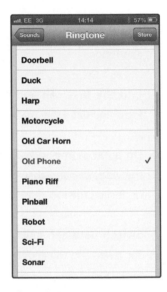

Above: The Ringtone setting screen allows you to preview and select which ringtone you would like. The selected tone will be highlighted with a tick next to it, as seen for 'Old Phone' above.

○ **Setting tones for calls, texts, emails and updates**: Sounds can also be assigned for incoming texts, emails and app updates. In order to edit these, head into Settings > Notifications; here you'll see the options to change the settings for all of your apps, including Messages and Phone.

○ **Assigning ringtones to contacts**: Go into Contacts and select the person for whom you want to set a new ringtone. Hit Edit > Assign Ringtone and choose from the list.

RECENT CALLS SCREEN

The Recent Calls feature on the iPhone lets you review calls you've made, received or missed and is accessed via the Recents icon at the bottom of your Home screen. This list of calls can be filtered to display All (every incoming and outgoing call whether answered or not) or Missed (incoming calls you didn't answer). From here, you can also access details such as when the call was made, how long it lasted and contact information for the person you called or who tried to call you.

Above: Swiping an entry in the Recent Calls list will display the Delete option in red. This entry can then be removed from your calling history.

○ **Call information:** Tap the small blue circle with the arrow to find out when a call was made, how long it lasted, when it started and finished, and any caller information stored in your Contacts.

○ **To return a missed call:** Simply tap the name of the caller and the phone will dial.

○ **To delete an entry from your Recents:** Swipe any entry in either direction and a red Delete button will appear. Hit the button and that item will be consigned to history.

Hot Tip
If one of the calls listed is from someone who isn't in your phonebook, it's easy to add them. Just tap the arrow pointing right, followed by **Create New Contact** and fill out their details. They will then show up in your **Contacts**.

VOICEMAIL

The days where the majority of us feared speaking into a machine are largely gone. Voicemail is now an essential tool for collecting those important – and not so important – messages while you're otherwise engaged. Some excellent new features on the iPhone now make it even easier to set up and manage this tool.

SETTING UP VOICEMAIL

The first time you tap Voicemail on your new iPhone, you will be prompted to create a password and to record a custom voicemail greeting. Getting set up should only take five minutes; here's how to get your voicemail up and running.

Above: Greeting screen with custom 'Record a greeting' option selected.

Recording a Personalized Greeting

The iPhone comes with a pre-recorded generic voicemail greeting, which is fine, but we recommend creating your own personal message for people to hear when you ignore their calls and send them to voicemail. Before you begin recording, it's worth making sure that you're in a nice quiet place and that you've planned what you're going to say. This should help you to get the perfect message in less than five takes.

○ **To record your greeting**: Select Voicemail from the Home screen and then tap Greeting, followed by Custom and Record. This will start the recording. Just speak your carefully worded greeting into the phone as if you were speaking on a phone call. Whether you go for wit or play it straight, it is entirely up to you but remember that it's not just your friends who'll hear it. Once you're done recording, press Stop.

○ **To review your greeting**:
You can review the greeting by pressing Play. If you're happy, hit Save or if you want another go follow the steps above again.

Set an Alert Sound for New Voicemail

You can assign a specific ringtone to let you know when someone leaves you a new voice message. Go to Settings > Sounds and then tap New Voicemail but remember that if the phone is set to silent it won't sound alerts.

Change Your Voicemail Password

Your Voicemail is password-protected so that you can keep all your secrets safe and avoid any information falling into the wrong ears, should you lose your phone or have it stolen. In order to set yours up go to Settings > Phone > Change Voicemail Password. This is also the password you'll use should you wish to dial in and collect voicemail from another phone.

Above: You can select a password to protect your voicemail messages.

Hot Tip

Quickly send an incoming call to voicemail by pressing the on/off button on the top of your iPhone twice. If you're using the headset you also can tap the microphone twice to send the caller directly to voicemail.

VISUAL VOICEMAIL

Forget wasting time listening to voicemails in the order they were left! Visual Voicemail now lets you see all the messages in your Voicemail inbox and select which you listen to first. No more wading through 10-minute monologues from family members when you really need that vital work update. This is not yet available on all carriers.

Above: The Visual Voicemail screen allows you to prioritise which messages you want to listen to or delete first.

How to Spot When You Have Voicemail

There are two ways to see when you have voicemail. On the Home screen, a notification will appear in the status bar along with a red circle over the Voicemail icon. This circle also shows the number of new voicemails awaiting your attention. The same red circle and number also appear over the Voicemail icon within the Phone app. In both cases, tapping will bring up your Voicemail list.

Visual Voicemail Explained

Below is a list of the main things you need to pay attention to while your using Visual Voicemail.

○ **Caller info**: In most cases the caller's name and phone number will appear but where no details are available, you'll see Unknown or Private Caller.

○ **The blue dot**: This signifies voicemails you've not yet listened to. If it doesn't have a blue dot that message has already been played back.

○ **Playing a voicemail**: Tap the name or number of the message you'd like to hear, followed by the Play/Pause button.

○ **Access contact info**: Tap the blue arrow that appears next to the caller's name and

number. From here you can also add this person to
your phonebook.

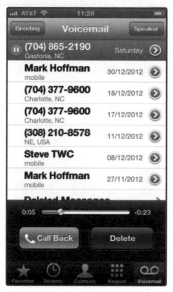

○ **Fast-forward and rewind messages**: Drag the scrub
bar to jump to interesting points within a message, just
like you would with an online video. Something from
the message you need to hear again and write down?
Just toggle back.

○ **Returning a call**: Press the green Call Back button.

○ **Deleting a message**: Hit Delete. Note that in some
cases it's possible to undelete a message if deleted by
mistake by scrolling to the end of the messages list
and selecting Deleted Messages. Then tap the message
and press Undelete.

Above: Playing a mesage from the
Visual Voicemail screen presents you
with the options to 'Call Back' or 'Delete.

Hot Tip

**The Phone app on the
iPhone does not provide a way
to save your voicemails in the
long term but you can download
third party apps like PhoneView
on to your PC or Mac. These
let you back up voice and
text messages.**

Retrieving Voicemail from Another Phone

If you don't have your phone with you,
or the phone has run out of battery,
you can still access your voicemails.
Either dial your own phone number
and follow the prompts (you'll need
your voicemail password) or call your
carrier's remote voicemail number.
In both cases you'll be back in a world
where you have to listen to messages
in sequence.

FACETIME

FaceTime (or video calling) is a great way to bring people together over long distances, letting you see fellow callers while you chat in real-time. It's as simple as making a regular call, although both callers must have an iPhone 4 (or newer), Mac or recent iPod Touch.

SETTING UP A FACETIME CALL: STEP BY STEP

1. Dial the person you want to FaceTime as if you were making a normal call (note: if you're FaceTiming someone on a Mac, iPad or iPod touch you'll need to use the person's email address to initiate contact).

2. Once the voice call is connected and you've decided to move to video, hit the FaceTime icon on the call screen. Decline [red] and Accept [green] buttons appear on the other caller's screen. Once they've accepted, expect a short delay before their face appears on your display.

3. In addition to the other person, you'll also see a smaller inset image of yourself on your phone screen. Use this to ensure you're showing the other caller what you want them to see – normal protocol dictates this should be your face but it's a free country.

Left: While on a Facetime call most of the screen will show the video image of the other caller, although there will be a smaller image of yourself shown at the top of your screen. The three icons at the bootom of the screen allow you to mute, end the call or switch between your iPhone's front and rear cameras.

4. Three additional icons appear on the phone screen while you're making a FaceTime call: Mute, End and Switch Cameras. The first two are self-explanatory; the latter lets you flick between the iPhone's front and rear cameras – handy if you want to show someone something else in the room.

Blocking FaceTime Calls

You'll always get the option to decline a FaceTime call but should you decide you never want to be contacted this way, you can disable it. Go into Settings, select Phone, look for the FaceTime section and ensure it is switched off. Once that's done, people calling you won't get the option of doing so via FaceTime (you can always turn this back on at a later date if you change your mind).

Above: When you receive an incoming call you may wish to select the FaceTime option. If you do not wish outgoing calls to display this option, switch FaceTime off in settings.

Flicking Between FaceTime and Apps

A big bonus of iPhone's multitasking skills is the ability to use apps while on a FaceTime call. It's great if you need to check diary dates, grab information from email or find something on a map. To do this, tap the Home button while the FaceTime call is live and then navigate your phone as normal. Once you decide to return to your FaceTime's mugshot, simply tap the green bar at the top of the screen.

Hot Tip

With the arrival of iOS 6 you can now use FaceTime over your 3G network but make sure you check your data plan first or you could end up with a hefty bill.

SMS

SMS (Short Message Service), or text messaging, has become one of the nation's most popular forms of communication, with an average of five texts per day sent for every person living in the UK in 2011. In this section we'll show you how to master the art of texting from your iPhone.

SENDING AN SMS

Sending text messages on the iPhone is pain-free but there are many different ways to achieve your desired effect. These depend on whether you're writing a new text to a new contact, replying to somebody's message after receiving an alert or browsing a past chat and getting the urge to text from within a conversation.

Above: The message screen allows you to send new messages, respond to messages and edit your message inbox.

○ **To send a new message:** Tap the Messages icon on your Home screen to jump into the Messages app. Here you'll see a list of all your received messages and a pencil and paper icon in the top right corner. Hit that icon and type in a contact number, start typing a name or hit the blue + button to choose someone from your address book. Write your message in the field below and then hit Send.

○ **To respond to a message from an alert:** Hit Reply to bring up the conversation view, type your witty retort into the Text Message field and hit Send.

○ **Texting multiple people:** If you want to send an SMS to more than one person – let's say to share a change of location for dinner – all you have to do is tap a second time in the

To field and select a second contact from your address book. Repeat this for everyone you wish to contact.

- **To forward a text**: Select the Edit button from within a conversation, tap the bubbles containing the content you want to forward until red check marks in circles appear to indicate those you've chosen. Next, hit the blue Forward button, add a recipient and share away.

- **Failed texts**: From time to time you'll attempt to send text messages when there is no network coverage. You can still write these and hit Send; they will show up as unsent messages in the conversation view. To send them you'll have to enter the conversation again when you regain network access.

Above: Use the text forwarding option by selecting the content you wish to forward, entering recipients and pressing send.

Hot Tip

Want a speedier way to type texts? There are lots of apps, such as Swype, that replace the iPhone's keypad with smart input technology that can predict words as you swipe over letters rather than tapping each one.

RECEIVING SMS

Receiving and responding to incoming texts will be one of the activities you do most with your new iPhone. People can get irritated if you don't give them an instant response to their message. Whether or not you subscribe to this outlook is your call but in this section we'll show you how to spot incoming messages, leaving you free to decide how quickly you reply.

Alerts and Notifications

Your iPhone can let you know you've received an SMS by trumpeting out a sound, vibrating in your pocket, popping up with on-screen alerts, or all or none of the above. You'll soon find which works best for you (in other words, the one that annoys you and those around you the least).

Setting Up Your Alerts and Notifications

In order to manage your SMS alerts and notifications, go into Settings > Notifications > Messages; here you'll see a number of editable options to customize your experience. From here you can manage whether – and how many – text alerts appear in the iPhone's Notification Centre, which on-screen alerts you'd like use, set your text alert tones and dictate how often alerts should repeat.

You can also decide if alerts should be allowed to show up on the Lock screen or whether everyone who sends you a text should have the same rights as the special people in your Contacts.

Spotting Incoming Texts: Types of Alert

- ○ **Banners:** These appear briefly across the top of your screen and display the person who sent the message along with the first line of text from the SMS (this is optional). After a few seconds, on-screen Banners disappear automatically.

- ○ **Alerts:** These appear in the middle of the screen. Again, they show who the message is from and an optional line of the message. However, you'll need to choose to either Close or Reply to these to make the Alert disappear.

Right: Home screen displaying text alert in the centre.

Above: The messages app icon displays the number of unread messages in your inbox in a red badge.

○ **Badge:** Just like Mail, Voicemail and Reminders, you can set your phone to display the number of unread messages over the Messages app icon.

○ **Tones:** You can assign one notification sound for all new text messages or choose a tone for a specific contact. If you want a general noise alert, go to Settings > Sounds > Text Tone and choose one of the listed alerts.

To Preview or Not to Preview?

You can choose whether you want the first line of a new text message to show up on your screen within Banners and Alerts. If you're worried about privacy – there are some things you don't want to broadcast to those around you – we suggest switching this off in the Notifications settings section.

Above: The Text Tone screen allows you to assign an alert tone to incoming text messages. You can also assign contact-specific tones. Selected tones are highlighted and ticked, as shown.

Hot Tip

To assign a special sound alert to let you know when a very important contact has texted you, go into Contacts, select the contact, hit Edit > Assign Text Tone and then choose your sound from the list.

MMS

MMS (Multimedia Messaging Service) is the ability to send messages that contain photos or videos. Sending MMS messages is as straightforward as sending a text but it's worth noting that video and photo messages will use extra data so it's worth checking your data plan before you start firing video clips across the globe.

SENDING AN MMS

Once you've mastered text messaging (see page 86), you are only a couple of steps away from being able to enhance your messages with photos and videos via MMS.

Sending Multimedia Messages

Follow the steps you would to send a normal text. Tap the Messages icon on your Home screen to open the Messages app, hit the pencil and paper icon in the top right corner and add a contact. Then hit the small camera icon that appears in the bottom left-hand side. You can then choose whether to add an existing video or picture from your gallery or to take a new one. Once you've made your selection, you can choose to write some text to accompany it; then hit Send.

Sending Photos and Video from Your Gallery

You can MMS photos and video while browsing through your gallery. When viewing an image, tap on the icon that shows an arrow on a screen (appearing on the bottom left) and this will bring up the options to share via your social networks, email and MMS. Just select MMS, choose a contact and hit Send.

Above: Selecting the camera icon on the compose message screen presents you with different options.

RECEIVING AN MMS

Viewing MMS messages is no different than reading an incoming text. The notifications and alerts follow exactly the same principles and you'll see photos and videos you've been sent come up within the text conversation view. Small thumbnails appear in bubbles alongside normal SMS messages and a simple tap on the picture or video will blow them up to be viewable full-screen.

If you like what you see you'll also be able to download these photos and videos to your phone's Camera Roll by tapping and selecting Save.

Forwarding an MMS

When it comes to forwarding content, MMS messages are treated just like texts. While you're in a conversation, click Edit and then tap the bubbles containing the content you want to forward until red check marks in circles appear to indicate those you've chosen. Next, hit the blue Forward button, add a recipient and share away.

Right: Selecting the Send a Photo option from the Camera Roll view, this displays a choice of functions, allowing you to share an image using social networks, email or MMS. If selecting MMS, just choose a contact and send away.

Hot Tip

You can save images from the web and send them via MMS. In the Safari browser, touch and hold an image, and you'll be given the option to save it to your Camera Roll or copy it to paste into an MMS or email.

GETTING CONNECTED

WAYS TO CONNECT

If you wish to access the internet, you'll need to be connected to either a Wi-Fi network or a mobile internet network. The iPhone is capable of accessing both.

Above: The Wi-Fi settings screen shows which Wi-Fi networks are available.

WI-FI

Wi-Fi is the wireless technology that allows you to connect to the web. The iPhone features the Wi-Fi technology similar to your computer or laptop, meaning you can easily access the web without physically plugging into the network.

Connecting to a Wi-Fi Network

In chapter one we ran through connecting to a Wi-Fi network while setting up the phone (see page 26); however, here's how to connect to a new Wi-Fi hotspot.

○ Press the **Settings** button and select **Wi-Fi**. If the Wi-Fi switch is set to off, toggle it on.

○ The iPhone will scan your locale for available **Wi-Fi networks**. After a few seconds you'll see available networks listed under the Choose a Network header.

○ **Select the network** and, if it isn't password-protected, you'll be connected. A blue tick will appear next to the network and the **Wi-Fi icon** will show in the title bar.

○ If the network is password-protected, you'll be taken to an **Enter Password** screen. Type this on the keyboard (letters will be obscured by dots after you type). Once complete, press **Join**.

Hot Tip
Wi-Fi passwords are often 'case sensitive', meaning that capital letters should be typed as such.

You should now be able to browse the internet, download apps, and send and receive email. As a test, return to the Home screen, enter the Safari app and attempt to load a webpage (see page 99). If you're in a public hotspot, then further log-on information may be required.

Above: Loading webpages in Safari displays them on your screen, allowing you to scroll to read text or view images.

Remembering Your Wi-Fi Places

The great thing about using Wi-Fi is that once you're connected to a network, it will remember you next time you're within range. So, if you connect to your home network, the iPhone will automatically pick up the signal and register your phone on the network when you get within range. The same applies at your local café, bar or coffee shop. In some places, you may be required to re-enter a password.

Ask to Join Networks

If you don't wish to automatically log on to a Wi-Fi network whenever you're within range, enter Settings > Wi-Fi and toggle the Ask To Join Network switch to on. You'll then receive a notification informing you that a known network is available.

Hot Tip
When you're in familiar places, make sure you're automatically connected to Wi-Fi so you can save on mobile data.

Wi-Fi Problems

Connecting to a new Wi-Fi network, especially if it's a public place, isn't always straightforward. Often you'll see a notification claiming that you've entered an 'Incorrect password'.

○ The iPhone keyboard can be quite **fiddly** and it's easy to make **mistakes**, so try again.

Above: Entering an incorrect password will prevent you from joining a Wi-Fi network.

⊙ There may also be 'case' issues so check with the person who gave you the password to ensure that **upper** and **lower case** letters are correct.

⊙ Even if the password you've entered is accepted, the location may be **experiencing problems** that prevent you from connecting to the internet.

⊙ **Check** with the venue to ensure that everything is working correctly.

MOBILE INTERNET

Although the world is slowly moving in that direction, not everywhere you go will offer access to Wi-Fi. Thankfully, the iPhone offers advanced mobile data connectivity, which provides over-the-air access to the internet in the same way that you're able to make calls and receive texts.

Mobile Internet Allowances

When you sign up for a mobile contract, you'll be given a mobile data allowance by your network. Cheaper monthly tariffs will only offer around 250MB to 500MB of data, which may be enough for light users. More expensive tariffs will offer 1GB, 2GB or even unlimited mobile data, which are more suited for heavy internet users who like to download music, stream video and download lots of applications.

Hot Tip

When deciding on which mobile data tariff to select, ask your network how much data you've used on average per month previously, but bear in mind that you'll probably be using more on the iPhone.

Types of Mobile Internet

Depending on where you are in the country, the speed of mobile internet varies a lot. Just like with regular mobile signal, in urban environments you're likely to receive faster and more reliable internet connectivity than you would in rural areas.

To that end, the current speed of your mobile internet will be reflected by the following letters and symbols in the iPhone title bar (*see* page 30).

- **4G LTE:** The most advanced mobile internet speeds are just starting to roll out across the UK. To access these speeds you'll need to sign a special data contract.

- **3G:** The most common third-generation internet speeds, suitable for browsing the web, using social networks, watching videos, and sending and receiving email.

- **E:** If you're still using the first iPhone you'll only have access to basic 2G speeds on the EDGE network.

- **Circle icon:** If you see a circle icon in place of any of the above, you have no mobile data connectivity.

Connecting to Mobile Internet

Once the iPhone is activated, it will automatically connect to mobile data networks wherever they are available. Unlike Wi-Fi, you'll never need passwords to access these, as the data comes as part of your mobile contract.

Configuring Mobile Internet

While you will be connected automatically to the mobile internet, you can still turn it off or control which apps use mobile connectivity.

Above: Although automatically connected, you can choose to turn off the connection to your mobile internet data by toggling from On to Off.

○ Enter **Settings > General** and then hit **Mobile Data**.

○ Here you can toggle the **Mobile Internet** switch to **On** or **Off**.

○ **Turning off** mobile data means that browsing the web, sending/receiving emails and downloading apps can only be achieved over **Wi-Fi**.

Mobile Data Exceptions

Within the Settings > General > Mobile Data screen you can control whether some features are accessible using mobile data. This could be important if you have a limited data contract. You can toggle the on/off switches for features like iTunes, which concerns whether you can download music and movies over 3G.

Preserving Your Mobile Data Allowance

As explained above, most mobile contracts now offer limited data allowances per month. Here are five ways to ensure you don't incur fees by going over your allowance.

1. Let Wi-Fi do the heavy lifting: If you're downloading movies or music for a long trip, be sensible and download large files while you're at home.

2. Log on to Wi-Fi networks wherever possible.

3. From the Settings > General > Mobile Data menu you can control which apps can use mobile data; you can toggle iCloud, iTunes, FaceTime and more.

4. Close apps when you're not using them. Sometimes apps can use a lot of data by checking for updates when you're not using them (see page 160).

5. Control how often individual apps search for new data (see page 160).

USING THE INTERNET

Now you're connected, via Wi-Fi or mobile data, you can start using your iPhone to access the internet through a web browser.

SAFARI

Safari is the web browser built in on the Apple iPhone. It is a mobile version of the desktop software that has traditionally been popular on Mac and PC computers. As it is one of the most used apps on the device, it sits neatly within the dock on the Home screen (*see* page 30). The icon is represented by a blue compass.

Loading a Web Page

In order to load a web page, hit the Safari icon on your Home screen. You should see a blank white page with a blue border at the top of the screen and another at the bottom. If you've used Safari before, the previous page will be displayed.

Above: Safari is the default web browser on the iPhone and sits in the dock.

1. Hit the address bar at the top left corner of the screen which reads 'Go to this Address'. This will summon the keyboard.

2. Type in an internet address (e.g. bbc.co.uk, apple.com).

3. Press the blue Go button to load the web page.

Hot Tip

There is no need to type the www in the address bar; simply the name of the site and the suffix (e.g. .com, .co.uk) will take you to the page.

Above: Typing a search term into the Safari search bar will bring up previous related searches.

Searching the Web

When you open Safari, you'll notice the small Search bar to the right of the address bar. Tapping within this will expand the Search bar and bring up a page denoting your previous searches. Start typing a search term (i.e. BBC Sport, Buy an iPhone) and press Search to load the Google Search results.

Hot Tip

There's a dedicated Google Search app that's free to download from the App Store, which offers a better, more comprehensive searching experience.

Finding Information Using Siri

What good is a personal assistant if it can't help you find the information you need to know? Ask Siri a question and there's a good chance it'll find an answer from the internet (the search provider Wolfram Alpha to be precise) within seconds – and here's how.

1. Hold down the Home button for a second until the Siri microphone pops up.

2. Tell Siri your question, i.e. 'How tall is the world's tallest man?'

3. Siri will say 'I'm on it', 'Checking My Sources' or 'Let Me Check'. If it can find an answer, it will load the information on a notepad, as seen in the screenshot to the right.

Above: Siri wil respond to your question by rapidly searching the internet for relevant information and loading the answer on a notepad.

Searching Google/Wikipedia Using Siri

If Siri cannot find an answer to your question (for example, 'How many number ones did The Beatles have?'), it will ask, 'Would you like me to search the web?' and you can hit the on-screen prompt or say 'Yes'. This will load Safari and Google Search results for your query. Alternatively, you can just say 'Search Google for...'. This command also works for Wikipedia.

NAVIGATING A WEB PAGE

Once you've loaded your favourite page, you can use the iPhone's responsive multi-touch screen to accomplish everything you could when using a keyboard and mouse on a PC or laptop – and a whole lot more.

Browsing a Web Page

If you're reading a news story on a web page, it's unlikely that all the text will fit on the iPhone's screen. You can browse a web page 'below the fold' by scrolling the page up, down, left or right with your finger.

Above: You can use your touchscreen to zoom in closer on a web page.

Zooming into a Web Page

Unless the web page you're visiting has been optimized for viewing on a mobile screen – as has been the case for many sites – it can be quite difficult to read text and get a close look at photos without squinting. Here's how to zoom in on specific areas of the page.

- **Pinpoint the area** of the page you'd like to **zoom on** and place two fingers on the screen.

- **Push outwards** with both fingers until you've zoomed sufficiently and then **let go** of the screen.

The second method is to quickly double-tap a specific area of the display, which will offer a precise zoom.

Clicking New Links

You can open an individual link on a web page by tapping a headline, category header or a picture. The new link will load in the current window, replacing the existing page.

Hot Tip

To remove all the clutter from a web page (design, pictures, links) and make it easier to read the article, just click the Reader button in the address bar – only the text will remain.

Hot Tip

In order to open a link from a web page in a new window, press the link and hold it to bring up a dialogue box. Then select Open in New Page.

Moving Back and Forth

Quite often, when browsing the internet, you may wish to take a step back, perhaps returning to the website's homepage to read a different story or going back to Google search results to look at alternative sites. To do this, you can hit the Back arrow in the bottom left corner of the screen. Likewise, you can use the adjacent Forward arrow to travel in the other direction.

Refreshing a Web Page

If you're using Safari to check the football scores or waiting for concert tickets to go on sale, you can refresh the page to display the most up to date information by hitting the circular arrow in the address bar, thus reloading the current page.

Mobile-optimized Websites

As more and more people are accessing the internet on smartphones and tablets, websites are changing their look and feel to ensure they're suitable for reading on devices like the iPhone. If a site has been optimized for mobile it will automatically redirect to the mobile site when you type in the address (for example, bbc.co.uk becomes bbc.co.uk/mobile). Also, such a site will not require any zooming and content will be neatly arranged horizontally for easy scrolling.

Above: The BBC mobile website is specifically designed for mobile scrolling.

Opening a New Web Window

If you'd like to open a new web page without leaving the one you're on, you can easily do so by pressing the two windows in the bottom right corner of the Safari page. This will launch a thumbnail image of the current web page and at the foot of the page you'll see an icon, which says New Page. Clicking this will launch a new blank page while preserving the first page; you can have up to nine windows open at any one time.

Above: To close a web page, select the window icon in the right corner of Safari, then click on the red cross, as above.

Navigating Between Multiple Web Windows

When multiple windows are present, the icon in the bottom right corner of the screen will feature a number representing how many windows are open. In order to switch between them, hit the number and swipe left or right between the windows, touching the thumbnail to maximize that page.

Closing a Web Window

If you have finished reading a web page and want to close it down to minimize some of the clutter, select the window icon in the bottom right corner of Safari and click the red cross in the left corner of the thumbnail.

Adding and Accessing Bookmarks

Bookmarks allow you to store webpages for easier access in the future. It could be a favourite website or perhaps a link to a product on a shopping site like Amazon. This is what you need to do to save a page to your Safari bookmarks.

1. Hit the Share icon directly above the Home button.

2. Select the Bookmark icon from the pop-up Share screen. From here, you can give the page a new save or select a specific folder.

Above: Clicking the Share icon enables you to bookmark favourite websites and add them to folders.

3. Press Done to add the page to your bookmarks.

4. In order to access your Bookmarks, select the open book Icon at the foot of the Safari window.

5. Tap an address to open that page in the existing window.

Hot Tip

If there's a website you use every single day, you can hit **Share > Add to Home Screen** to add an icon to your desktop which will take you directly there.

Sharing Web Pages

Easy sharing of web content is increasingly important to smartphone users and Safari excels at this. You'll see the Share icon (the box with an arrow leaping from it) all over the place when using the iPhone, but here are the options when you select it within Safari.

- **Mail**: Selecting this icon will launch the Mail client with the web link copied into a new email.

Above: By selecting the Mail icon from the Share screen, your Mail account will be launched with the web link copied into a new email.

- **Message**: Same as above, but in the Message client.

- **Twitter**: Share the link with your Twitter followers.

- **Followers**: Post the link to your Facebook wall.

- **Add to Home Screen**: Adds an icon to the homepage for easy access in future.

- **Print**: If your iPhone is configured with a wireless printer you can print the contents of the web page.

- **Copy**: Copy the link to be shared elsewhere (into a document, a Skype conversation, alternative email app, etc.). In order to then paste the link, hold your finger down in a text box and select Paste.

- **Bookmark**: Saves the page to your Bookmarks (*see* page 104).

- **Add to Reading List**: Safari offers a handy Reading List tool, which lives within the Bookmarks. It allows you to put together a list of pages you'd like to read at a more opportune time. The Reading List screen is separated into two categories: All items and Unread items.

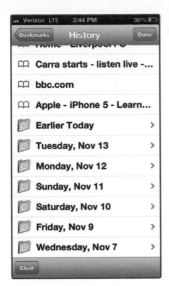

Browsing History

Bookmarks and Reading List are just a couple of ways to access familiar pages. Hitting the Bookmarks icon will give you the chance to select a History Menu, which stores information about all of the pages you've visited in the past week.

Also, when you start typing a website into the address bar, Safari will begin to predict the page you wish to access, based on your previous activity (for example, if you start typing Goo... then a list of activities related to Google may appear below). If you see your preferred site, simply select it from the list.

Other Web Browsers

Although Safari is the default browser and works extremely well in tandem with the rest of the iPhone, there are other options available from the App Store. Here are three of them.

○ **Chrome:** Google's Chrome is now the most popular web browser in the world. You can download a free mobile version for the iPhone, which features a 'tabbed browsing' feature.

○ **Opera Mini:** Another browser which is free to download and may be familiar to computer users; it promises faster loading times and mobile data saving.

○ **Dolphin:** A free, gesture-based web browser, which also lets you search the internet with your voice.

Above: You can download the mobile version of Google Chrome to use the 'tabbed browsing' feature.

SOCIAL NETWORKING

The web continues to become more and more social. Services like Twitter and Facebook are among the most important ways of keeping in touch with friends and our favourite celebrities, and of keeping track of what's happening in the world.

FACEBOOK

Facebook now has over a billion users around the world – and the chances are you're one of them. There are a number of ways to access the social network using your iPhone so let's explore them.

Facebook: Now built-in to the iPhone

One of the big changes with the new Apple iOS 6 software (available for iPhone 3GS, 4 and 4S models, and built-in to the iPhone 5) was deep Facebook integration. This means that you can share web pages, photos and videos, and post status updates directly to Facebook from within various apps on your phone.

Setting up Facebook

In order to link your Facebook account to your iPhone and enable all of the neat sharing functionality mentioned above, take the following steps.

1. Press the Settings icon, scroll down and select Facebook.

2. Enter your username and password into the fields provided, and then click Sign In.

Hot Tip

Don't have a Facebook account? Go to Settings > Facebook > Create New Account and follow the steps.

3. The next screen explains how Facebook will interact with your phone. Hit Sign In in the top right if you agree.

4. Once your username and password are accepted, Facebook will be synced with your iPhone.

5. Next, you'll get a notification asking you to 'install the free Facebook app'. You can do this now or later. Selecting Install will prompt a request for your Apple ID password.

Posting Status Updates

You can post directly to your Facebook wall from the Notification Center by dragging it down (see page 45) and selecting Tap to Post. This launches a unique Facebook dialogue box where you can type a status update and then select Post when you're done.

Above: The Facebook settings screen allows you to customize the way the app interacts with other iPhone apps such as calendar and contacts.

Sharing Photos and Videos

Once you've synced your Facebook account with the iPhone (see above), it's easy to post photos and videos to your Facebook wall.

1. Enter the Photos app and select Camera Roll.

2. Scroll up or down to find the photo of your choosing and click the Share icon.

Above: Select the Share icon on the bottom left of the Camera Roll screen to share a photo via your Facebook account.

3. Select Facebook to launch the
    picture or video within a Facebook
    dialogue box.

4. If you wish, add a message to
    accompany the item and press 'Post'.

5. The photo or video will upload in front
    of your eyes and soon appear on your
    Facebook wall.

> ## Hot Tip
> In order to instantly share a
> photo to Facebook, take the
> photo, hit the thumbnail in the
> bottom left corner to view it
> in the Camera Roll and repeat
> steps 2–5 above.

Sharing Web Links on Facebook
As we mentioned earlier in the chapter, Safari makes it easy to share web pages through the
Share icon at the bottom of the page. There's a dedicated Facebook button that enables you
to share interesting links with your circle.

Sharing to Facebook in Other Apps
As you download new apps (*see* page 157), you'll notice that a lot of them also incorporate
the Share icon that allows you to share to Facebook. The YouTube app (*see* page 178) is a
prime example. Other apps, such as Google Chrome, feature their own mechanism for sharing,
which also allows items to be posted to Facebook.

Share to Facebook (or Twitter) Using Siri
Rather than typing your Facebook message, you can have Siri do it for you.

1. Hold down the Home button to launch Siri and say, 'Post to Facebook.'

2. Siri will ask, 'What would you like to say?' and a blank Facebook post will load.

3. Dictate your status update, i.e. 'I'm looking forward to the new Foo Fighters album'.

4. The text will appear in the screen and Siri will say 'I updated your Facebook status. Ready to post it?

5. You can say Post, Cancel or hit those commands on the screen.

Facebook Events in Your Calendar

We'll explain more about getting the most from the Calendar app later in this chapter but, by syncing your iPhone with your Facebook account, event invites and friends' birthdays will be incorporated into your calendar, which means they'll also appear in the Notification Center. It's a great way to ensure you never forget another birthday.

Above: You can use Siri to update your Facebook status.

Above: If you have synced your Facebook account with your iPhone, your events will appear listed in your calendar.

Facebook Contacts in Your Address Book

Integrating your Facebook account will automatically download your friends to your Contacts Book, along with any information they may have shared on Facebook (e.g. phone number, email address, birthday). In order to add your Facebook friends to your Contacts, go to Settings > Facebook, turn Contacts to on and then select Update All Contacts.

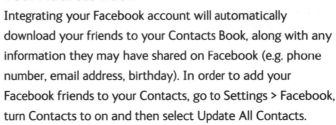

Hot Tip

Adding anything from 100 to 1,000 people may make your contacts book a cluttered mess. In order to remove them, head to Settings > Facebook and toggle Contacts to off. The same applies if you don't want Facebook interfering with your calendar.

Installing the Facebook App

While you can access Facebook through the Safari browser by heading to Facebook.com, it's far more efficient to do so via the self-contained Facebook app. You may have installed this when syncing your phone with Facebook (see page 108). If so, a blue 'F' icon will appear on your desktop. If you haven't installed it yet, access Settings > Facebook and hit Install at the top of the menu. You will be asked for your iTunes password; hit OK and the app will install.

Using the Facebook App

The Facebook for iPhone app allows you to perform most of the tasks you're used to on the main site – here's a summary of the main features.

- **News feed**: Launch the app and you'll see the News Feed with the latest posts from your friends. Tapping the posts will allow you to comment or 'Like', whereas tapping a name will take you directly to their page

Above: Access your Facebook messages through the blue Facebook alert bar.

- **Status/Photo**: Hitting these commands allows you to post a status update or a photo or video from your Camera Roll.

- **Check In**: If you'd like to alert friends to your current location (i.e. home, a favourite restaurant, sporting event), hit Check In and select from the list of nearby places.

- **Notifications**: The blue Facebook alert bar will display incoming friend requests, messages and notifications from friends.

- **Chat**: Hit the chat icon in the top right-hand corner of the app to load a menu listing your friends. Click a name to start a chat conversation.

○ **Profile page**: To access your own profile page, hit the menu icon in the top left corner and select your name.

Controlling Facebook Notifications

With messages, likes, comments, photo tags and friend requests flooding into most of our Facebook accounts, you may not want those Notifications to be delivered to your phone. Select Settings > Notifications > Facebook to control whether alerts appear in your Notifications Center, on the Lock Screen and determine the type of notification you'll receive (None, Alerts, Banners, etc.).

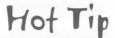

Hot Tip

If you have a notification that requires your attention, a number will appear next to the Facebook (or Twitter) icon on your Home screen.

Above: You can turn off and customize your Facebook notifications through the Facebook settings screen.

TWITTER

For those unacquainted with Twitter, it is a hugely successful and free to use social network that enables its 300m+ user base to post updates of 140 characters or less. Users can 'follow' and interact with their friends and favourite celebrities using the site.

Twitter Integration With iPhone

Twitter enjoyed system-wide deep integration with the iPhone when Apple launched iOS 5 in 2011. It's easy to update your Twitter feed, to share photos, videos and web links, and also to receive notifications without even entering the Twitter app.

Setting up Twitter

Setting up Twitter integration is a very similar process to enabling Facebook, which we explained on page 108. Head to Settings > Twitter, insert your username and password in the required fields and press Sign In. As with Facebook, you'll be encouraged to install the free app, which can be done immediately or later. If you don't have an account, you can select 'Create New Account' and follow the on-screen steps.

Above: The Twitter settings screen provides you with the option of downloading the free Twitter App.

Hot Tip

The features explained reside within the newest version of the Twitter app, which is continually evolving. To ensure you have the latest version, select App Store > Updates.

Sending a Tweet

Once you've entered your Twitter details or registered for an account through your iPhone, you can start sharing your 140-character pearls of wisdom with the world. In order to send a tweet, simply drag down the Notifications Center menu and hit the Tap to Tweet button. This will summon a new Tweet box and the iPhone's keyboard. Simply type your message – remember to take note of the character limit ticker that informs you of how many characters you have left – and press Send to post it to your Twitter account.

Mentioning Other Users in Tweets

To 'mention' a friend (or a user you follow) within your tweet, type the @ symbol on the keyboard and begin typing their username (e.g. @DavidCameron). As you type, the names of people you follow will appear to match your keystrokes. When other users are 'mentioned' in tweets, they will be notified.

Sharing Photos and Web Pages on Twitter

This is largely the same as the Facebook sharing explained on page 109.

○ **To tweet a photo**: Use the Share icon in the Camera Roll and select Twitter.

○ **To tweet a web page**: Use the Share icon in Safari and select Twitter to load a new Tweet, complete with a shortened version of the web link.

Both of these methods will summon a new Tweet box, as mentioned above. As well as the embedded link, photo or video, you can add your own accompanying message, mention other users and add a geo-location.

Above: You can share photos via Twitter by selecting the Share icon in the Camera Roll. You can then add an accompanying tweet.

The Twitter App

The iPhone's built-in Twitter functionality allows you to send out Tweets; however, in order to read other people's posts, receive notifications, reply to Tweets and direct messages, and control who you follow, you'll need the free Twitter app.

Installing the Twitter App

If you declined to install the Twitter app when setting up the account (see page 114), it's easy to rectify. Just go to Settings > Twitter and hit the Install icon at the top of the screen. You may be asked to enter your Apple ID password, but after that it will be installed on your device and the Twitter icon will appear on your Home screen.

Using the Twitter App

As you can see from the screenshot overleaf, the Home section of the app is dominated by the feed of posts made by people you follow. They're usually a combination of friends, celebrities, companies or news organizations. Scroll up and down the screen to view more Tweets and tap one to interact with it. From here you can reply, 'retweet' (i.e. share the message with your followers) and add it to favourites.

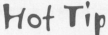

Above: The Twitter app displays your Twitter feed in an easily readable and scrollable format.

Sending a Tweet With the Twitter App

These are the steps to follow in order to send your own tweets.

1. Hit the paper and pen icon in the top right corner and begin typing.

2. Click the Photo icon to take/choose a photo to attach.

3. Hit the compass arrow to log your location.

Above: It is easy to add a new tweet using the Twitter app.

4. Touch the @ symbol and type usernames to mention other users.

5. Hit # symbol to add a hashtag and contribute to a topic.

6. When your tweet is complete, like the one below, press send.

Navigating the Twitter App

The Twitter app has four sections listed in the navigation bar at the bottom of the screen, as described below.

Hot Tip

In order to refresh your Twitter feed and load the newest posts, place your finger near the top of the screen and pull down.

- **Home**: The main page, this displays your Twitter feed of posts from those you follow. Pressing Home again will take you to the top of your feed.

- **Connect**: A list of replies and mentions made by other users in relation to your tweets.

- **Discover**: A snapshot of what your friends are up to, a list of 'trending' (popular) discussion topics and the chance to search for other users.

- **Me**: Your profile page, complete with your own tweets, profile pictures and the opportunity to send direct messages to friends.

Hot Tip

If you're adverse to the thousands of replies your celebrity status merits, you can turn everything off through Settings > Notifications > Twitter.

Managing Twitter Notifications

Once you've installed the Twitter app, you'll receive notifications when another user replies to your tweets, retweets or favourites one of your postings, or sends you a direct message. These notifications can be configured and opened in the same way as Facebook notifications.

OTHER SOCIAL NETWORKS

While Facebook and Twitter are undoubtedly the big two, the App Store offers dedicated portals to access some of the other more popular social networks (for a detailed guide to downloading apps, see chapter four).

○ **LinkedIn:** This professional social network allows you to list your CV and to connect with colleagues and potential employees.

○ **Google+:** Google has launched a neat app for its fledgling Twitter/Facebook rival, but so far users have been slow to engage.

○ **Pinterest:** An alternate social network that simply allows users to 'pin' picture-based objects from around the web to their own 'boards'. It's great for creative types looking to collate ideas, shopaholics and lovers of funny cat pictures.

○ **Foursquare:** This location-based social network encourages users to 'check in' at their favourite places in order to earn rewards and leave 'tips' for future patrons. Check in enough times and you could become the 'mayor' of your local trendy coffee shop.

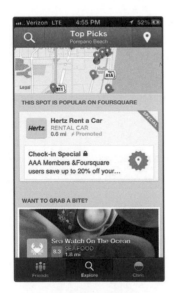

Above: Use the Foursquare app to check in at various locations and earn rewards.

EMAIL

In this section we'll talk you through the basics of setting up your various email accounts and the intricacies of sending and receiving email using the iPhone's Mail app.

THE MAIL APP

Email is, unsurprisingly, one of the iPhone's absolutely key pieces of functionality and most users will use it several times a day. As such, Apple has taken the liberty of placing the Mail app in the dock on the Home screen, meaning it's always easily accessible. Select the app and let's get you set up.

Above: With an iPhone, sending and receiving email is simple: you will find the Mail app in the dock of your Home screen (bottom right here).

Above: Select your preferred email provider from the Mail app set up screen.

SETTING UP YOUR EMAIL USING MAIL

Whether you're a Gmail, Hotmail, Yahoo Mail or Microsoft Exchange user, the Mail app has got you covered. When you first select Mail, it will ask you to set up an account and you'll see the screenshot below. Simply select the provider of your choice to move to the next step.

Configuring Microsoft Exchange

The iPhone is increasingly becoming the smartphone of choice for business users and, as such, it's easy to get your work email set up on the device, providing your company is happy to send you mobile data.

1. Select Microsoft Exchange from 'Welcome to Mail'.

2. Enter your work email address and the password you use to access it.

3. Fill in the description field ('Work' would make sense) and press Next.

Above: Access work emails from your iPhone by setting up Exchange.

4. Once the iPhone has verified the login information, you'll need some information from your IT guys to complete the setup. Ask them for the Server information (e.g. server.company.com) and the username you use for the account. Once you've entered that information into the respective fields, press Next.

5. If you've done this correctly, you'll see ticks appear next to all of the fields.

6. The next screen features the opportunity to add information from your account to the Contacts and Calendars applications. There's more on this in the Calendars and Contacts sections (*see* pages 47 and 56).

7. Once you're happy with this, press Save.

8. Within seconds, you should see a batch of emails arrive in your inbox.

Configuring Gmail, Hotmail, Yahoo and Others

If adding Microsoft Exchange email details seemed a little tricky and convoluted, don't worry because setting up your personal email is a doddle.

1. Select your email provider from the list on the 'Welcome to Mail' screen.

2. If you've added a different account already, you'll need to select Settings > Mail, Contacts, Calendar > Add Account.

3. To configure the likes of Gmail, Yahoo, Hotmail and others, add your name, email address, password and a description (optional), and press Next.

4. If you've entered the details correctly, ticks will appear next to all of the fields and the iPhone will progress to the final setup screen.

5. Once that's dealt with, press Save to start receiving emails.

Syncing Email With Other Apps

Depending on which email provider you're using, you'll be able to pull in details to work with some of the other apps on the iPhone. For example, Yahoo users can sync their Contacts and Calendars, as well as their Reminders and Notes (*see* chapter one).

Above: Once you have configured your account to the Mail app, you can sync details such as contact lists.

Combined and Multiple Inboxes

Once you've added your accounts, you can access them through the Mailboxes screen within the Mail app (see screenshot below).

Here you'll see the option to select All Inboxes, which will show you all emails in the same inbox, regardless of which of your configured email addresses they were sent to. You can also select each inbox individually to see segregated Exchange, Gmail, Yahoo, etc. accounts.

Email VIPs

The final option on the Mailboxes screen is to view emails from your VIPs (bosses, best friends, significant other, etc.). Select VIP from the Mailboxes and choose a specific contact from the Contacts app. These VIPs will have their own specific inbox and you can easily configure special notification settings for their mails by using the VIP Alerts button.

Above: The Mailboxes screen allows you to view your accounts individually or through a combined inbox.

Hot Tip

If you've received an email from someone to whom you'd like to give VIP status, click the name of the sender (highlighted in blue in the 'From' field) and select 'Add to VIP'.

DEALING WITH EMAIL

Sending and receiving email using the Mail app is just as easy as text messaging.

Sending Email

Here's a quick step-by-step guide to firing off your first email.

1. Open the Mail app and select the New Message icon in the bottom right corner.

2. Choose your recipient by typing the address into the
'To' field (known contacts will be suggested as you type);
for example, joe.bloggs@email.com.

3. To select a recipient from your Contacts press the
+ icon and select from the list.

4. Click Subject and type an email subject
(e.g. Dinner tomorrow?).

5. Tap the main body of the email, which says
Sent from my iPhone, to begin typing.

6. Once you've finished typing, press
Send in the top right corner. You'll
hear a 'whoosh' sound once the
email is sent.

Attaching Pictures or Video

Here is how to include a picture or video in your email.

1. Hold your finger down anywhere within the body text field to
summon a pop-up menu that says Select, Select all, Paste.

2. Press the arrow to the right of that menu and select
Insert Photo or Video.

3. Touch a photo from the Camera Roll to launch a preview and
press Choose if you're happy. This will attach the photo to the
email. Repeat these steps to add multiple photos.

Above: Attach a photo to your email by
holding your finger in the main section
and selecting 'Insert Photo or Video'.

4. Touch the screen to place the cursor above or below the photo to continue typing your email.

Sending an Email Using Siri

As with Facebook, Twitter and Messages, you can dictate an email to Siri to save typing it out.

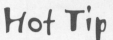

In order to launch a new email with the photo already attached, select Photos > Camera Roll and choose the photo. Hit the Share icon and select Mail from the pop-up list.

Above: Siri can be used to send emails to people listed in your contacts.

1. Launch Siri and say 'Send an email to...', including your recipient's name. Siri will load an applicable person from your Contacts app. If you have many email addresses for the same name, it'll ask you to choose (i.e. Home, Work).

2. A cool postcard will launch. Then you can dictate a subject, body copy and say 'Send'.

3. Siri will attach a Mail stamp to the postcard and send the email. Simple.

Multitasking With Mail

As with all iPhone apps, you return to an app at the precise place where you left it. So, if you need to leave the Mail app while composing an email, you can safely browse to another app without losing your draft. When you return to the application, your half-composed email will still be waiting for you.

Saving Drafts

Alternatively, you can easily save draft emails until you return to them later. When composing an email, press Cancel in the top left corner of the screen. From the pop-up menu, you can choose Delete Draft to discard, Save Draft to add it to the Drafts folder or Cancel to continue composing the email. Naturally, once the drafts are saved, they will live in the Drafts folder within your email account (*see* Accessing Mail Folders on page 128).

Receiving Push Email

The default setting for the Mail app is to bring in new emails as they arrive. This is called 'Push email' and is extremely handy. It is designed so that it will continually go to ask the server if there are new emails, rather than the user having to refresh manually. When you receive a new email, you'll usually be notified by a sound (the 'ding' sound is the default) and a visual alert. These can be configured using Settings > Notifications > Mail. A number next to the Mail icon on the Home screen will also evidence new emails.

Fetch Email

While Push email arrives automatically, Fetch email instructs the Mail app to check for new arrivals at regular intervals. Select Settings > Mail > Contacts > Calendar > Fetch New Data. Turn Push off and select from the options under Fetch. You can choose every 15, 30 and 60 minutes or Manually. Regardless of these settings, every time you open the Mail app, it will automatically check for new email.

Above: You can choose whether you receive your emails by Push or Fetch. Choosing the Fetch option saves battery.

Scheduling Email Updates

If your work email is configured with your iPhone, come 5 pm you may not want emails from colleagues coming to your device but would still like to be notified of

Hot Tip

If you find that having your work email and your personal email coming to the same app is a little too much to handle, why not download the dedicated Gmail, Yahoo Mail and Hotmail apps from the App Store to help segregate work and pleasure?

received personal emails. To that end, you can change the scheduling settings for each account. Here's what to do to schedule when the Mail app searches for new mails.

1. Select the Settings app and choose Mail, Contacts, Calendars.

2. Select Fetch New Data and then Advanced.

3. Choose the email account in question and select Manual. This will only bring in new email when you manually enter the Mail app.

4. When it's time to go to work in the morning, follow the steps above and change the schedule to Push.

Refreshing Email

Regardless of the schedule you've requested for email updates (see above), you can manually load new emails at any time. From any of the menu pages (Accounts, Inbox, Sent, etc.) you can refresh by placing your finger on the screen and pulling down. At the foot of the page you'll see a message telling you when the folder was last updated (e.g. Updated 13/11/12/ 12:35 PM).

Reading Email

When you open the Mail app and select your Inbox, you'll see all emails listed with the sender, subject, first line and sent time. Any unread emails will be represented by a blue dot to the left of the message.

Touching the preview will open the email, full screen, on your iPhone, thus allowing you to read the full contents. Hitting the Inbox arrow in the top left navigation bar will take you back there, whereas hitting the up or down arrows will take you to the previous/next email in the list.

Selecting Web Content from Email

Many emails will feature web links, perhaps from work colleagues, friends or shopping sites that require you to click to view more content. You can tap a hyperlinked picture or a line of text and the webpage will instantly open within the Safari browser.

Replying, Forwarding and Managing Emails

After reading an email, you can use the menu at the bottom of the screen to decide what to do next. The blue menu at the foot of the app gives you several options, as shown in the screenshot below.

- **Flag**: Add a flag to the email to signal its importance or mark it as unread.

- **Folder**: Click this icon to move the email to a new folder (e.g. Receipts, Invoices, Trash).

- **Bin/Archive**: Discard the email by removing it from your inbox and sending it to the trash or your archives.

- **Reply**: The arrow allows you to Reply, Reply All (if there's more than one sender/recipient), Forward it to a new contact or Print the email (providing that your iPhone is configured with a Wi-Fi printer).

- **New Message**: From anywhere within the Mail app you can hit the New Message icon to begin typing an email.

Above: After selecting the arrow icon from the bottom of an email, the reply screen will be displayed. You can then choose to Reply, Forward or Print.

Deleting Emails

A junk-free email account would surely bring a state of inner peace that the twenty-first century human has yet to achieve. Therefore, there are bound to be plenty of emails that you don't wish to read or store. Here's what you need to do to dispose of them.

1. Select Edit from within your inbox and an empty circle will appear next to all emails. Touching this circle will highlight it with a red tick.

2. Highlight all of the emails you'd like to delete.

3. Press the red Delete button at the bottom of the screen.

4. From here you can also Move emails to new folders and Mark them for further attention.

Conversation View

If you've received multiple emails from the same sender on the same subject (if you've replied and they've replied, and so on...), you'll see a number in a square box to the right of the message preview. Selecting these emails will bring up a new page with previews of all of the emails associated with that conversation.

Accessing Mail Folders, Drafts and Trash

Not all of your email activities take place within your main Inbox. There will be times when you'll want to access your Sent Mail, Draft emails and perhaps emails within your Trash that were deleted prematurely. From the Mailboxes page select from the Accounts menu. This will take you to a list of Folders where you can access Drafts, Sent Mail, Spam, Trash and more.

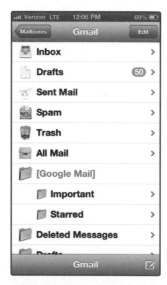

Above: The Folders screen allows you to organise and easily access different aspects of your emails.

CALENDAR

Back in chapter one, we introduced the iPhone's built-in Calendar app; however, in order to make the best use of it, it's better to have all of your email and social networking accounts set up, which we have achieved in the previous pages.

THE CALENDAR APP

This app isn't hard to find. It sits on your iPhone's Home screen and looks like one of those neat single-page desktop calendars. Touch the app to see a monthly grid view, displaying the month, the year and an icon for every day in the month. Today's date will be highlighted in blue, with any events (e.g. a birthday, listed below the grid).

Above: The Calendar app will display today's date on the Home screen (here it is second from right, third row down).

Above: Touching a date in the calendar app brings up details of appointments entered for that day.

Navigating Around the Calendar App

Touching another date icon in the monthly view will take you directly to that day and show the appointments you have listed then. Touching an event will provide more details, such as the time, location, invitees and more. At the foot of the app, you'll see the following five navigation buttons.

Above: Calendar list view shows the date and time of upcoming events.

○ **List**: This shows a list of the upcoming events, providing the date and time for each.

○ **Day**: Selecting this will show an hour-by-hour summary of everything happening on that day.

Hot Tip

When viewing the calendar in monthly view, days with events scheduled will have a black dot underneath the date.

○ **Month**: This takes you to the monthly grid view, with each day represented by an icon. You can use the arrows on either side of the month (for example, 'November 2012') to move back and forth.

○ **Download**: This icon shows any invitations yet to be added to your calendar.

○ **Today**: The button on the far left automatically takes you back to Today, presented in List, Day or Month format.

Configuring Multiple Calendars

By default, the Calendar app features All Calendars. This means that it will feature events from all the accounts you've given the Calendar app permission to access. When setting up your email, you would have been asked whether you wanted it to sync with the Calendar app (see page 121). In order to select which accounts send events to the Calendar app, select Settings > Mail, Contacts, Calendar then choose an account (e.g. iCloud, Gmail, Exchange) and switch the Calendar button to on or off.

Adding Facebook Events to the Calendar

As Facebook notifies you of your friends' birthdays and also keeps you abreast of events you've been invited to, it can be a good idea to enable the Calendar so you're notified on your smartphone. Providing you've already added your Facebook account (see page 108), you can select Settings > Facebook and then allow the Calendar to use your account by toggling the switch to on.

iCloud Calendars

One of the coolest and most useful iCloud features is the Calendar functionality, which syncs across multiple Apple devices. Therefore, if you make an entry on your iPhone, you'll see it on your Mac computer, iPad and even on the web-based iCloud.com page. To enable the iCloud Calendar go to Settings > iCloud and turn Calendars on.

Above: Using the iCloud calendar is a great way to view and manage your appointments on your Mac, and can be synced across devices.

Viewing Multiple Calendars

If you've configured multiple email accounts and allowed Facebook to push events to your iPhone, you can choose to view them all or just one at a time. From the app's main screen select Calendars in the top left corner. Pressing the various items in the list (e.g. iCloud Home) will add/remove a tick, making it visible in the All Calendars screen.

Searching Your Calendar

With all of these accounts pushing data to and from the iPhone app, it can be difficult to keep track. To search for details of a certain event, you can use the Search Calendars bar near the top of the app; tap and type to see events.

ADDING TO THE CALENDAR

The iPhone Calendar app will automatically sync with your various email and Facebook accounts (more on that later), but you can also add items manually – here's how.

1. Open the Calendar app and hit the plus button in the top right corner.

2. From the new Add Event screen, enter the Name and Location of the event in the fields provided.

3. Select the Starts/Ends/Time Zone menu from the Add Event screen.

4. With Start highlighted in blue, select the date and time using the scroll reels. Select End and do the same.

Left: Add an event to your calendar by using the keyboard to enter the title, location and times. You can also choose to receive an alert.

5. If this is an all-day event, toggle the switch to on.

6. If your meeting is fortunate/unfortunate enough to be in a different time zone, select Time Zone, delete the current location from the City Name field and begin typing the new location. Press Done in the top right corner.

7. Select Calendar from the Add Event screen to choose within which of your calendars the event will appear (e.g. iCloud, Exchange).

8. Select Done in the Add Event screen to add the event.

9. Browse to the date of the event to see it residing within the Calendar.

10. Tap the event to see the Event Details.

Above: You can schedule an event using Siri, who will add it to your calendar.

Adding a Calendar Event Using Siri

Most of us aren't lucky enough to have a secretary to arrange our affairs, but thanks to Siri, everyone can feel like a big shot. Part of Siri's remit is to manage your calendar, meaning that you can add appointments to your Calendar app and invite attendees in a couple of seconds. For example, you can launch Siri and say, 'Schedule meeting with Joe Bloggs at Starbucks at 5 pm on Friday.'

If you confirm this meeting with Siri, it will be added to your Calendar app and an invitation will be sent to your recipient. You can also cancel appointments in the same way by saying, 'Cancel appointment with...'

Setting an Alert for a Calendar Event

The settings on the previous page will allow you to add the essential details for a new event, but there's a host more options to customize. For example, any Calendar app worth its salt will alert you when the event is approaching. When adding or editing (select the event and press Edit) an event, you can customize when you will receive an alert and select anything from None to 2 days before. Press Done to confirm (you can also add a second alert).

Calendar Alerts in Notification Center

Calendar events are automatically configured to appear in your Notification Center, which can be accessed at any time by pulling it down from the top of the screen. This means that the events for that day will always appear as a constant reminder throughout the day.

Selecting Calendar Alert Styles

Above we explained how to set timers for alerts but from within Settings > Notifications > Alert Style you can select Banners or Alerts. The latter requires your attention before doing anything else, so selecting this gives you a greater chance of seeing the notifications. You can also toggle whether the alert will appear in the Lock screen, and select the Vibration style and Alert Tones.

Invite Contacts to Calendar Events

If you're planning a business meeting or a party, you can use the Calendar app to invite potential attendees. From the Add/ Edit Event screen, select Invitees; this will then load the Add Invitees screen, the iPhone's keyboard and a To: field.

Hot Tip

Every time you update the event (time, date, notes), invitees will be notified by email.

○ In order to add an **invitee**, you can start typing the email address in the To: field. Alternatively, press the **+ icon** to load your contacts.

○ Selecting a **Contact** will add them to the list.

○ Select '**Done**' to send out the invites.

In order to see who has accepted, rejected or yet to reply to the invitation, tap the event to access the Event Details screen.

Scheduling a Regular Calendar Event

If you have to go to the same meeting every week, you can schedule a repeating Calendar event. From the Add/Edit Event screen, select the Repeat option. The default setting is Never, but you can choose Every Day, Week, Two Weeks, Month and Year (great for remembering your grandmother's birthday!).

Above: You can invite contacts to calendar events by selecting Add Invitees and then typing email addresses or adding them from your contacts.

Adding Notes to Calendar Events

The Notes section of the Add/Edit Event allows you to add details about the event, which is great for yourself and folks you've invited. Tap the Notes field and then use the iPhone's keyboard to type details; press 'Done' when you've finished.

Deleting an Event

Despite your best-laid plans, meetings get cancelled and parties get postponed. So, to delete the event from your own Calendar (and those who may have accepted invites), select the Edit button from the Event Details page, scroll down to the bottom of the screen and hit Delete Event.

Above: In Calendar Day view you move an event by dragging the event bubble up or down manually.

Moving/Extending an Event

You can change the time, date and duration of an event by using the Edit screen, but there's an easier way too. Select the Day view in your Calendar and place your finger on the event bubble.

You'll then be able to drag the event to earlier or later in the day by moving it up and down. To switch days, drag it to the left of right. When you've settled on a place, let go of the bubble.

You'll also notice markers at the top and bottom of the bubble; dragging these in or out will change the length of the appointment.

Hot Tip

In Day view, turn the iPhone on its side to view multiple days at a time.

OTHER CALENDAR APPS?

As with the other key apps discussed in this section (Safari and Mail), the built-in Calendar is designed to play nicely with the rest of the phone. It is beautifully designed and it will easily accommodate Microsoft Exchange, Gmail, Yahoo, Hotmail and any other accounts you have set up on your phone. There are lesser third-party apps on the App Store, but there's no real reason to look elsewhere.

LOCATING YOURSELF

Whether you're searching for directions, finding a restaurant nearby or checking-in at your favourite venue, you'll use the iPhone's mapping and location services more than you think. Over the next few pages we'll explain how to get the best out of your new personal GPS device.

LOCATION SERVICES

When you first set up your iPhone (see page 25), you will have been asked whether you wanted to enable Location Services and allow certain apps to access your geographical location. This isn't just handy for Maps but also for social networks, weather applications, shopping and ticketing apps, the camera, and Siri.

Above: You can switch on Location Services in your privacy settings.

How Location Services Works

The iPhone is a clever device. It can triangulate your approximate location by using the GPS (satellite) signal, your proximity to a mobile network tower and data from the Wi-Fi network you're logged on to. When Location Services is in operation, you'll see a compass arrow in the iPhones title bar.

Enabling Location Services

If you switched on Location Services when setting up the phone, you're good to go; otherwise, it's easy to rectify. Head to Settings > Privacy and switch Location Services to On.

Enabling Location Services for Individual Apps

On the Settings > Privacy > Location Services screen you'll also see a list of apps which are

Above: The Location Services menu allows you to view and edit which apps are utilising this service.

requesting access to your location. There are a few obvious ones, such as Maps, Compass and Weather, but also other less obvious ones, like Camera (in case you were wondering, this is for the purpose of geo-tagging the locations of your photos). Toggle the switches to On or Off for each app, depending on your preference.

APPLE MAPS APP

If you're using the latest version of the iPhone software (i.e. iOS 6) then you'll have a brand-new Maps app, built by Apple to replace Google Maps (see page 145). The built-in Maps app brings voice-controlled, turn-by-turn navigation to the iPhone, replacing the need for a dedicated sat nav.

Opening Maps

The first time you open Maps, a graphical map will load with a blue dot surrounded by a circle, which represents your current location. If you're on the move, this blue dot moves with you – pretty cool, huh?

> ## Hot Tip
> To pinpoint your Current Location at any time, press the compass arrow in the bottom left corner.

Navigating Around Maps

You can move around the Maps app screen using a lot of the same gestures we've encountered in apps like Safari. Here's a quick refresher.

- **Scan**: You can move around the map by placing your finger on the screen and moving it in any direction.

- **Zoom**: You can zoom in and out on any map by touching the screen with two fingers and moving them in or out. Double-tapping the screen with one finger will also zoom.

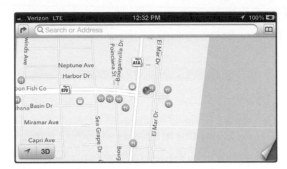

Above: You can choose to view your map in landscape by turning your phone on its side.

○ **Accelerometer**: Turn your phone on its side to see the map in landscape view.

Changing the Map

As well as the standard graphical view, you can also access satellite imagery. At the bottom right corner of every map is a page corner; lifting it (as seen in the screenshot right) gives the opportunity to switch the view, show traffic information, drop a pin and even print the map.

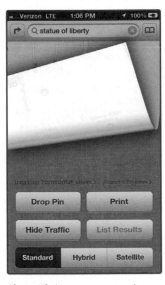

Above: Lift the page corner to select other map view options.

Above: You can view your location and directions in Satellite view.

SEARCHING FOR A LOCATION

There are a number of ways to search for a particular place (e.g. a restaurant, office, friend's house) if you're using the Maps app. You'll find each of them in a tab above your map. In the next few pages, we'll explain how to use the Directions, Search and Bookmarks options.

Get Directions from the Maps App

Selecting the Directions arrow from the main Maps screen will load a new screen allowing you to navigate from a Start to an End point.

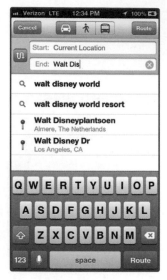

Above: Enter your destination and touch Route to dispay directions.

1. Select your mode of transport by tapping the car, walking or public transport icons at the top of the screen.

2. The Start field will display Current Location. In all likelihood, you won't be changing that.

3. The End field is blank. Tap within this to type your location. If that location is familiar, it will appear in the recent locations below the fields. If it is an unfamiliar location, suggestions will appear as you type. Touch one to select. If the destination does not appear, continue typing until the field is complete and then press Route.

5. A new Map screen will load, featuring your start point (represented by a green pin), your destination (represented by a red pin) and the route you'll need to take (represented by a blue line).

6. The menu bar will feature potential routes (touch one on the map to change it), while it'll also give you an estimate of how long it will take and how far away the destination is.

Hot Tip

Hit the flip route button to the left of the Start/End fields to reverse your route – perfect if you're navigating the same route in reverse.

7. You can use the blue dot to navigate your own way *or* press the blue Start button to launch voice-controlled, turn-by-turn navigation.

Above: The Maps screen will pinpoint your initial and final locations and display a route line between them.

Finding a Place

Selecting the Search or Address bar within the main screen in the Maps app is probably the simplest way to find what you're looking for.

1. Tap the bar and begin typing your request. It can be an address or the name of a place, venue, restaurant, etc.

2. Familiar locations and suggestions based on what you've typed will appear below the search bar. Tap a suggestion or continue typing and then press Search.

3. The nearest result (if you search for McDonald's there may be loads!) will appear in the centre of the map, accompanied by a red pin and along with an interactive bar displaying the name of the place.

Above: You can search a location or venue using the Maps search bar. The results will show as pins on the map.

4. In order to convert the location into Directions starting at your Current Location, tap the green car icon next to the name and then press Start to commence turn-by-turn navigation (*see* page 143).

5. For more details about the location, hit the blue arrow next to the name for a full address and phone number (if applicable), or to add to bookmarks, add to contacts or to share the location.

Left: Selecting the blue arrow seen in the screenshot above displays further information about the location, as seen here for McDonald's Restaurant.

Directions to Bookmarks, Recent, Contacts

To the right of the Search bar is the open book icon from which you can access locations already stored in your phone. There are three options here to choose from: Bookmarked locations, Recent items (both locations and routes) and Contacts (if they have a street address associated with them). Select any of these items to view them on the map and then press 'Start' to begin turn-by-turn navigation.

Dropping a Pin

If you're browsing around using the Maps app, it can often be easier to drop a pin on a location you intend to visit. In order to drop a pin (purple in colour), hold your finger down on the area of the map; an address bar will pop up, enabling you to get directions.

Above: Use Siri to help you find your way; simply launch Siri and ask for directions to where you want to go.

Get Directions Using Siri

Siri is neatly integrated with the Maps app to allow you to load directions within seconds. Launch Siri by holding down the Home button and request directions to a location; you can say 'Directions home' or 'Take me to the nearest petrol station' and the Maps app will instantly load with turn-by-turn navigation. This will save you about a minute of typing.

Left: When browsing using Maps you can drop a pin to view directions to a chosen location.

MAPS' TURN-BY-TURN NAVIGATION (iPHONE 4S AND IPHONE 5 ONLY)

In this section we have continually referred to the turn-by-turn navigation functionality. Just like an in-car sat nav unit (e.g. a TomTom or Garmin), it will guide you from your current location to your destination with a series of detailed audio and visual instructions.

Using Turn-by-Turn Navigation

As we've explained in the previous pages, there are multiple ways to access the turn-by-turn navigation feature within Maps, which works for both walking and driving. Press Start from any Directions page (*see* page 139) and follow these instructions.

Above: Turn-by-Turn navigation is accompanied by voice instructions.

1. Once you hit Start, the iPhone will issue its first voice command (i.e. in 300 feet turn left on to Castle Street), while a visual road sign will display the same message.

2. The street/road/motorway you're currently on will always be displayed in the tab at the top of the screen, along with the direction number (i.e. 1 of 10)

3. Your position will be illustrated by a moving compass icon, which will move as your car/feet do. As you get close to your next turn, the command will be repeated.

4. When you're on the next street, you'll be told how long to continue on that street for.

5. The next instruction will appear in visual form on the screen. Continue to follow these instructions until you reach your destination. Once you're at your destination, press End.

Hot Tip

iPhone's multitasking allows you to leave the app and still receive voice instructions. When you leave, the title bar will flash green with the message 'Touch to return to Navigation'.

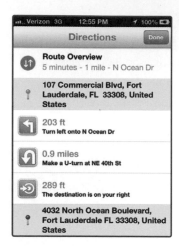

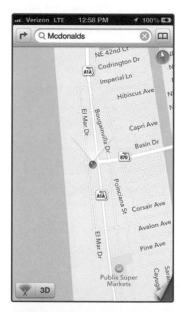

Above: You can also choose to opt for a written summary of turn-by-turn directions.

Viewing Written Directions

As cool as turn-by-turn navigation is, you may just need a written summary of the directions to make a mental note. Touch the text icon in the bottom left corner to see a written overview of every turn you need to take.

Enabling the Compass

While hitting the compass icon once in the bottom left corner will take you directly to your Current Location, hitting it twice in quick succession will actually load the fully functioning compass, turning the icon purple and loading a new NESW indicator in the top right screen.

Using the Compass

The compass can be great for getting your bearings before setting off on a walk or drive. As with any compass, when you change direction, it changes direction with you. You simply move the phone until the compass indicator is pointing directly in your required direction and off you go. Experiment with this and you'll soon get the hang of it.

3D Maps and Flyovers (iPhone 4S and iPhone 5 Only)

Pressing the 3D button in the bottom left corner of a Maps page will display a slanted, rather than a bird's eye, aerial view of the map, which can help sometimes to offer a greater perspective.

Above: The compass feature is a useful way of getting your bearings.

However, in some major cities, Apple has launched a Flyover feature, allowing you to zoom around the cityscape. Where it is available, the 3D button will be replaced by the skyscraper icon you see in the screenshot below.

OTHER MAPS OPTIONS

The launch of Apple Maps within iOS 6 caused quite a stir. Quite frankly, upon launch, it was a bit hit and miss, but it is getting better. Because of this, iPhone 5 users and those with older models who upgraded to iOS 6 have been turning to other options.

Above: Apple's flyover feature allows you to browse the cityscape for several major cities.

Google Maps

Google Maps had been the default Maps app on every iPhone until the iPhone 5, when Apple launched its own version to replace it, as explained in the last few pages. If you're using an iPhone 4 or 4S running iOS 5 then you'll still be able to access the popular (and superior) Google Maps app. If you've upgraded to iOS 6 then you'll notice that Google Maps is no longer available. Thankfully, Google launched a new version of the app, which is available to download from the App Store free of charge.

1. Select the App Store icon from your handset.

2. Select Search and type in Google Maps.

3. When the results shows up, hit Install and enter your Apple ID password.

4. The app will be downloaded to your **homescreen**.

Above: Apple Maps is the default app on iOS 6, but you can download a free Google Maps app from the App Store.

Apple Maps vs Google Maps

The experience of using Google Maps is very similar to using Apple Maps. If you're able to navigate around one, you'll find using the other quite easy indeed. Like Apple Maps, Google's app features turn-by-turn navigation, walking, driving and public transport directions. Google Maps may not have Apple's 3D flyover option, but at present its Maps are more accurate and feature more detailed points of interest than Apple's fledgling solution.

Traditional Sat Nav Providers

Afraid that smartphones were putting them out of business, the likes of TomTom and Garmin have launched iPhone apps with advanced features such as live traffic updates, regular map updates and advanced lane guidance. Search for Maps on the App Store to identify them (hint: you'll recognize them by the large price tag).

KEY USES OF LOCATION IN OTHER APPS

As explained earlier in this section, the Locations Services is useful beyond maps apps – here are some of our favourites.

Find My Friends

Some would say it borders on stalking, but Apple's Find My Friends app allows you to see where fellow iPhone users are on a map. Friends have to agree to follow each other's movements, so it's not like all of a sudden your precise location is available to everyone. To that end, it's great if you're meeting someone, as you can track each other's whereabouts. Here's how it works.

Above: The Find My Friends app allows you to locate and track fellow iPhone users on a map.

1. Open the App Store, search for Find My Friends and click Install to download this free app.

2. Sign in by using your Apple ID, accept the terms and conditions, and allow the app to access your location and contacts.

3. In the open app select the + icon in the top right corner, add a contact and press Send.

4. Once the recipient accepts the invitation, you'll see their name in your list of friends. Selecting their name will show their position on the map.

Hot Tip

Rather than permanently tracking each other, you can invite friends to temporarily share their locations. Open Find My Friends, select Temporary, choose a contact, a name for the event and a time for the location-sharing to end.

5. Press the compass to see your position in relation to theirs.

6. Select Message to send them a text – perhaps letting them know you're on the way.

7. Select Maps to open your respective locations in the Maps app and get directions (as explained on page 139).

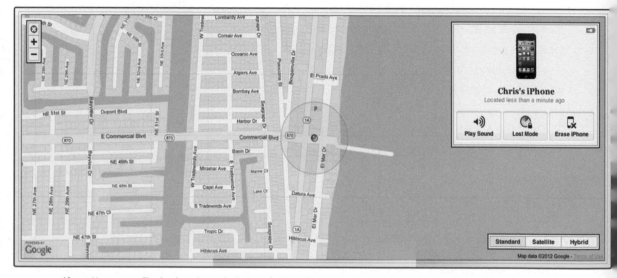

Above: You can use iCloud and another apple device to find your iPhone on a map, as well as performing other functions.

Find My Phone

When setting up the iPhone, you were asked to set up Find My iPhone (*see* page 29), which can be of assistance if you lose your device or it is stolen. If you didn't enable it during setup, go to Settings > Location Services > Find My iPhone to switch it on. From another device (i.e. an iPad, Mac computer or iCloud.com) you can locate the phone on a map, make it play a sound (if it's lost under the sofa cushions), send it a message (e.g. 'Oi, bring my iPhone back!'), or remotely lock or erase it to protect your data.

Above: Check in on the Foursquare app to interact with others and keep friends updated on your whereabouts.

Checking In

Apps such as Facebook and Foursquare (*see* page 118) and a host of others encourage you to use Location Services to 'check in' at a location. It's a cool way to keep track of places you've visited and perhaps interact with other people at the same venue.

Exploring an Area

There are lots of apps available from the App Store (Google+ Local, Yelp, Foursquare, etc.) that use location data to let you know what's in your area in terms of restaurants, hotels, attractions, petrol stations, cash machines and more.

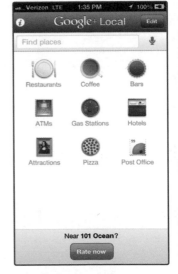

Above: Apps like Google+ Local are useful for finding places if you don't know the area.

Shopping

Applications like Groupon and LivingSocial use your location to find deals applicable to your locale.

Entertainment

Applications such as Movies by Flixster and Ticketmaster use your location to bring you details of events closest to you. Flixster will automatically display cinema show times closest to you, while Ticketmaster will automatically draw your attention to concerts in your area.

Geo-tagging Photos

We'll touch on this more in chapter five, but the iPhone's Camera app and popular third-party apps like Instagram are able to add a geographical location to your snaps. On the iPhone's Camera app the co-ordinates will be stored within the photo data.

APPS

BUILT-IN APPS

The iPhone comes with a host of built-in or 'native' apps. From Maps and Siri to Mail and FaceTime video calling, there are plenty of tools to make life easier. However, the real joy starts when you add your own from the App Store. This section shows you how to enhance your iPhone experience.

COMES WITH...

The iPhone 'native' apps provide a basic level of functions for the handset. These tools are the backbone of your iPhone user experience and you'll get to know them well as you become familiar with your device. Here's a brief description of the main ones.

- **Mail:** This is an Email app that lets you manage email from a range of mail providers centrally.

- **Contacts:** This app is your address book. It's a store for the phone numbers, addresses and contact information of all your friends, family and colleagues.

- **Passbook:** Designed to help manage boarding passes, coupons, tickets, gift cards and passes by keeping them all stored in a scannable digital format in the app.

- **Camera:** The basic Camera app works with the iPhone's built-in cameras to shoot stills and record HD video. A new panoramic mode has been added with iOS 6 and there are additional editing tools now included.

Above: The Passbook app helps you to store all your passes in one place.

- **Safari:** This is the iPhone's standard web browser. If you're an iPhone 5 user, this is optimized for the new 4-inch retina display. You can also auto-sync bookmarks and favourites across iOS devices.

- **FaceTime:** Apple's built-in video calling app that lets you chat face-to-face with other iOS and Mac users.

- **Music, Video and Photos apps:** These apps let you manage your multimedia content.

- **Game Centre:** A central store that keeps track of your high scores and achievements across numerous iPhone games.

- **Siri:** Apple's voice assistant. This app lets you issue instructions and tackle tasks using voice commands.

Above: The Game Centre app organises your iPhone games and records scores.

Apple's Apps

There are other notable Apple-created apps worth checking out in the App Store. Here are three we recommend.

- **Newsstand:** Like your virtual bookshop or newsagents, this app helps you to organize your magazine and newspaper app subscriptions and downloads, as well as discover new titles you might like to read.

- **iBooks:** Your gateway to a virtual bookshop. Read, buy and enjoy everything from classics to bestsellers.

- **Apple Remote:** Use this app to control your iTunes on your computer over Wi-Fi.

Above: Newsstand app stores your magazines and newspapers.

GET APPS

Although the iPhone has some excellent apps as standard, you'll definitely want to add your own. Before your start the voyage of app discovery, it's worth thinking about how you want to use your phone, how much space you have and how much you want to spend.

Your Apple ID

Before you can download any apps, you'll need an Apple ID and to be logged in. On your iPhone, go to Settings > Store and follow the prompts. While you're setting up your account, you can also choose on which devices your purchased apps will appear – if you have more than one iOS device – and also whether you want to enable app downloading over 3G. We'd advise sticking to Wi-Fi unless you've got a very lenient data plan. Remember that, should you wish to share your phone, you can use multiple Apple IDs for a single iPhone.

Find Apps

Just like restaurants, bars or cafés, finding great new apps can be a real buzz. The main way to find out what's new and popular is on your computer via iTunes or via the App Store on your phone. Alternatively, just search the web or turn to page 250 for our guide to the top 100 apps first.

Hot Tip

Not sure what to download? There are plenty of app discovery sites on the web, such as www.t3.com/appchart, which have done the hard work trawling and testing what's on the App Store.

APP STORE

You can buy and download apps on the move, from anywhere in the world, but before you can use the App Store on your iPhone, it must be connected to the internet.

CHOOSE YOUR APPS

Once you are connected to the internet, here is what you should do to browse the App Store.

1. Select the App Store icon on your Home screen to launch the App Store.

2. You'll see five buttons along the bottom: Featured, Charts, Genius, Search and Updates. Each represents a different way to discover new content.

○ **Featured**: Displays a mini version of the iTunes App Store.

○ **Charts**: Links you through to the Top Free, Top Paid and Top Grossing charts, showing the most popular apps of the moment.

Above: Selecting the Featured button displays a smaller selection of the apps available on the App Store.

○ **Genius**: Suggests apps that you might enjoy based on past app downloads.

○ **Search**: If you know the name of the app, you can tap it into the search bar.

○ **Updates**: Clicking this displays any available updates for apps you've already downloaded.

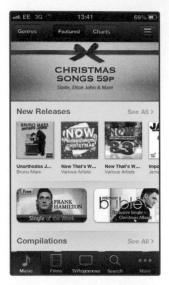

Above: You can open the App Store from iTunes.

iTunes

The iTunes App Store has been designed to be easy to navigate but here are a few pointers as to what each section contains.

○ **Opening the App Store in iTunes:** Launch iTunes on your computer, select iTunes Store from the Sources list on the left and click the Apps link. Next, select the iPhone tab and you'll see the iPhone section of the App Store.

○ **Navigating the iTunes App Store homepage:** The page is split into sections – shelves if you like – where the content changes regularly to remain topical.

○ **The Carousel:** The large image and the three related options in the centre at the top of the page all showcase themed app collections, which could be anything from iPhone 5-ready apps to suggestions for Christmas.

○ **Categories Drop-down:** In the top right-hand corner you'll find a drop-down menu of all the different app categories, such as Games, Books and Food & Drink.

○ **Top Paid Apps:** This section shows you the most popular paid-for apps in the App Store.

○ **Top Free Apps:** This section shows you the most popular free apps in the App Store.

○ **Top Grossing Apps:** This section displays the apps that have made the most money.

○ **Digging Deeper:** Each of the sections has a scroll bar for access to more recommended apps, plus a 'See All' button to take you to a list of all the apps in that category.

USE APPS

Once you've found the apps you want, you need to know how to get them into your iTunes, on to your iPhone and organized in a way that makes sense for you. In this section we'll show you how to download apps, create folders for easy navigation and also how to update your apps when new releases come along.

DOWNLOADING APPS ON YOUR iPHONE

In order to download an app on to your iPhone, tap the price button near the top of the screen – at this stage you may be asked to enter your iTunes Store account password.

Above: Screenshot of Angry Birds app in the App Store, indicating the price button and star rating.

The App Store will then close and an app icon will appear on your phone's Home screen. This will appear dimmed, with a blue progress bar next to it to indicate that it's downloading. Once the download is complete, the icon will appear sharp and the app is ready to use. Please remember that this app won't be in your iTunes library until you sync your phone.

DOWNLOADING APPS USING ITUNES

Within the iTunes App Store, once you've found the app you want, simply click the Get App or Buy App button. You'll be prompted to log into your iTunes account and when you're through security, the app will begin to download. Once it's downloaded, you'll need to sync your phone to make it appear on your handset.

ORGANIZE APPS

You can have up to 11 Home screens on your iPhone and – aside from the iPhone 5, which holds 20 – each of those pages can hold 16 apps or folders. Apps can be organized on the iPhone itself or via iTunes.

Move Apps

○ **Moving apps around on your iPhone**:

Press and hold any app icon until the apps being to wiggle. You can now move any of them on the screen: just press an icon and drag it to a new location. You can move it to another Home screen by dragging it to the far left or far right of the screen.

Above: By pressing and holding apps on the Home screen they will begin to 'wiggle' and can then be moved, deleted or stored in folders.

> ## Hot Tip
> It's worth checking the App Store regularly for special offers, as many app developers temporarily drop their prices to drive more downloads, particularly on Thursdays and Fridays.

○ **Moving apps around in iTunes**: Select your iPhone in the Devices section and choose the Apps tab. From here you can click and drag app icons to new locations on the screen. You can also change the order of your Home screens by dragging and dropping a screen upwards or downwards.

Create Folders

Creating folders is simple: press and hold any app on your Home screen until all the apps start to wiggle. Then press and drag the app you wish to drop into a folder on top of another app you also wish to have in that folder. If you're doing this in iTunes on your computer, just use the mouse to drag and drop the apps onto one another.

Multitasking: Switching Between Apps

If you're using iOS 4 or later then you can multitask, switching

from one app to another without closing them down; just hit your iPhone's Home button to return to your Home screen and choose the new app you want to use.

Web Clips

In addition to stuffing your Home screens full of apps, you can also add shortcuts to your favourite websites so that there's no need to open Safari. In order to do this, just open the site in Safari and click the + button, followed by Add to Home Screen.

Hot Tip
In order to access apps you've recently used, double-tap your iPhone's Home button to bring up the Multitasking bar. You can then swipe left or right to reveal apps with which you've recently played.

Web Apps

Unlike normal iPhone apps, which are pieces of software downloaded to your iPhone, web apps are essentially websites tailored specifically to work on the iPhone but still live out there on the internet. Generally, you need to be online but they have the bonus of not taking up storage space (you can find a full list of web apps at apple.com/webapps).

UPDATING APPS

App developers will issue updates of their apps in response to feedback they receive via iTunes or to release general improvements.

GET THE LATEST

Here are two ways to ensure you've got the latest version of all the apps on your phone.

- **On your computer using iTunes**: Select Apps in the iTunes library list on the left and click 'Check for Updates'.

- **On your phone**: If app updates are available, a circled number will appear above the App Store icon on your Home screen as well as on the Updates icon in the App Store. Within Updates, an Update button will appear next to the app that needs refreshing. You can either press that or hit the Update All button to do them all at once.

Remove Apps

Like most app-related tasks, you can either delete apps on your iPhone or on your computer iTunes – here's how to do both.

Above: A numbered circle on the updates icon indicates if updated versions are available for the apps that you already own.

- **In iTunes**: Click Apps in the source list and then click the app icon, followed by Edit > Delete.

- **On your iPhone**: Press and hold an app icon until all the apps begin to wiggle and then tap the X in the corner of the icon for the app you want to delete.

Above: Once you select an app to delete from wiggle mode you will be faced with a delete confirmation box.

○ **Permanent delete:** Removing an app using your iPhone doesn't delete it permanently. Unless you've also deleted it in iTunes, it will reappear on your phone the next time you sync it.

SHARE APPS

Sharing apps between your iOS devices will be taken care of automatically, provided you're syncing with iCloud. You can also select which apps you want to share to specific devices from within iTunes (*see* page 40 for our guide to setting up iTunes for more information).

App Settings and Preferences

The apps that come with the iPhone and the ones you download from the App Store have settings and preferences that you can edit. Every app will have its own functions available for customization and there are two ways to tweak them.

○ **In Settings:** Go to Settings and scroll down the page until you see a list of apps. Not all apps will appear here but those with settings to amend will. Tap the app name to take a look at what can be customized.

○ **In the app:** Look for the words Options or Settings, or alternatively, an icon that looks like a cog. Failing that, there may be settings hiding within a More menu.

Hot Tip

Even if you delete an app on your iPhone and in iTunes, it's not gone forever. Apple knows which apps you've previously purchased and these can be re-downloaded for free at a later stage.

MULTIMEDIA

CAMERA

Your iPhone is a pretty nifty compact camera, too. After reading the next few pages, you'll be convinced that you can leave your clunky compact at home.

STILLS CAMERA

The stills camera was once considered the iPhone's weak spot, but Apple has worked hard and improved it significantly over the years so that now there are few better smartphone cameras than the iPhone 4S and the iPhone 5. It will allow you to take 8-megapxel pictures good enough to blow up and place on your mantelpiece with pride.

Above: You can access the camera app from your lock screen by touching the camera icon and swiping upwards.

Taking Pictures
Once you have selected the Camera icon on the Home screen, you'll see the world in front of you on the iPhone's screen. This means that you're ready to start taking pictures, with one touch of the on-screen, in-app camera button. Simply touch it once to capture the image in a fraction of a second.

Taking Pictures With Volume Buttons
The iPhone has never had a physical camera button but in newer models you can use the volume keys as a shutter trigger, rather than the touchscreen. It can make it much easier to hold the camera steadily, rather than straining to reach the on-screen camera button.

Taking Photos Faster
Photographs capture a fleeting, never-to-be-repeated moment but by the time you've woken up the screen, swiped to unlock and entered a

pin code – and then opened the Camera app – that moment may have passed. Apple realized this and built easy access to the Camera app from the Lock screen. Swipe this icon up to open the Camera app directly from the Lock screen so that you'll be able to take pictures within a couple of seconds rather than 30.

Focusing in on Your Subjects

When you open the iPhone Camera app, the sensor will seek to focus automatically on the item in the centre of the frame and to give you a nicely balanced photo. You'll see a flashing blue square as the focus adjusts. It's wise to give the camera a second or so to complete this task before taking the photo.

Changing the Focus

Beyond autofocus, you can also tap anywhere on the screen to choose the precise area of the frame on which the sensor will

focus. This is called tap-to-focus. Tapping this will define a smaller focus area, which comes in handy if you'd prefer to concentrate on a specific item within the frame or even something in the background.

Above: You can tap anywhere on the screen, to bring up a blue outline and make that area the focus of your photo.

Face Detection

Naturally, you'll be taking pictures of friends and family (or yourself). When the iPhone spots faces in a picture, it will outline them with a green box, meaning that extra focus is being placed on them.

Left: The green box indicates face detection.

Getting Closer to Your Subjects

On most cameras there are dedicated buttons for zooming in and out on subjects, but this isn't the case with the iPhone where any zooming needs to be done on-screen. In the same way you'd zoom in and out on a web page, use your thumb and forefinger to pinch in and out.

> # Hot Tip
> Rather than maxing out the zoom and ending up with out-of-focus photos, it's better to be cautious with the zooming and crop the photo later.

Using the Flash

The camera on newer iPhone models (the 4S and 5 in particular) is very effective in low-light conditions and will produce better pictures than you'd expect. However, there are times when using the flash is unavoidable. You'll see an icon in the top left corner which says Auto, meaning that the iPhone will employ the flash at its discretion. You can tap this to override and turn the flash on or off.

Flipping the View

While the overriding purpose of the front-facing camera is for video chat (*see* page 84), it's also great for self-portraits and group shots when the cameraman is also in the frame. Touch the flip camera icon in the top right corner of the screen and take a photo as normal. The front-facing camera is not as good as the one on the rear but can still take decent snaps and removes the guesswork from self-portraits.

Other Camera Options

Apple has kept it simple with the iPhone: there are no 'scene' settings, such as sunset, action or sepia, like you see on some smartphones or cameras. There are just three choices within the Options menu at the top of the screen.

Left: The iPhone camera options allow you to make adjustments to your default camera settings.

Above: The iPhone camera is capable of taking high quality photographs, even when it has zoomed in on the image.

○ **HDR:** This stands for High Dynamic Range. Switching this on can offer much better photos in some circumstances. In simple terms, it takes a photo at three exposure settings – Underexposed, Overexposed and Normal – and combines the best elements of all three. When using this mode, a copy of the original photo will also be saved.

○ **Grid:** Turning the grid on offers a visual guide to help you frame your photos better.

○ **Panorama:** *See* page 168 for more information.

Right: You can turn on the grid option to bring up a visual guide which will help with framing your photos.

Move iPhone continuously when taking a Panorama.

Step 3: Once you press the camera button, the first shot of the panorama appears in the indicator.

DONE

Step 4: The arrow and the blue line indicate vertical movement, helping you to keep your panorama even.

Taking Panoramas (iPhone 4S and 5 Only)

The Panorama option is a new addition with the iOS 6 software and is perfect for taking wide-angled shots of landscapes, almost by recording the photo as a video clip. Here is what to do in order to take a Panorama shot.

1. Open the Camera app and select Options at the top of the screen.

> ### Hot Tip
> It's also possible to take earth-to-sky panoramas by turning the device on its side.

2. Select Panorama and hold the phone upright (panoramas cannot be taken in landscape).

3. Position the phone at the left edge of where you'd like the panorama to begin and press the camera trigger button. The first shot in the sequence will appear in the indicator.

4. As steadily as possible (use a tripod if you have one), pan the camera to the right, trying to keep the arrow in the centre of the spirit level-type indicator. The progress of the scene will be depicted within the indicator.

5. When you've completed the pan (you can move around 240 degrees), press Done.

6. In order to look at the results, press the thumbnail in the bottom left corner.

Step 6: The completed panorama allows you to capture wide-angled scenes.

RECORDING VIDEO

Any smartphone worth its salt is now capable of recording high definition video now. The iPhone 4S and iPhone 5 can record full HD video which matches what's on offer from some dedicated camcorders.

Camera to
Video Switch

How to Record Video

There is no dedicated icon for the iPhone's video camera; instead, it's easily accessible from a switch within the Camera app. You'll see the switch in the bottom right corner (top right corner when shooting in landscape mode); drag it to the video icon and press the red/grey button to start/stop recording. A timer logs how long you've been recording.

Above: When your iPhone is in landscape mode, the icon used to switch to video camera is located in the top right corner.

Take Stills While Recording Video

Even if you're shooting video of a gleeful youngster blowing out his birthday candles, the iPhone's camera is capable of performing double duty by capturing still shots. After pressing the record button, you'll see the stills camera icon pop up next to it. Touch this to save a still shot to your Camera Roll. Best of all, the shutter won't make its usual clicking sound and disturb your video.

Lighting Your Video Subjects

You'll notice on the video capture screen that the flash options are still present. When shooting video, the flash turns into a light, which allows you to still shoot decent quality video in the dark.

STORING/VIEWING PHOTOS AND VIDEOS

After you've taken a photo or recorded a video, you'll notice that it'll leap into thumbnail icon in the bottom left corner. This is your Camera Roll, where all of your recordings are automatically stored on your iPhone's internal memory. This is accessible in many different ways.

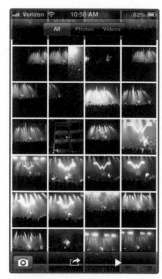

1. Press the thumbnail from within the photo capture screen to take you to the last shot. This is great if you're looking to share or edit a picture instantly.

2. Swipe left from the camera screen to perform the same functionality as above.

3. Select the Photos app from the Home screen to access the Camera Roll (featuring thumbnails of all of your photos and videos) and other albums you may have created (see the next page). Touch a thumbnail to begin interacting with it.

Above: Camera Roll shows you thumbnail images of photos and videos. Click on a thumbnail to edit or share.

Backing up Photos

As we explained in chapter one, your precious memories are safeguarded whenever you back up your iPhone – physically via iTunes or over the web via iCloud (*see* page 28 for backup instructions).

Storing Photos With iPhoto (Mac Users Only)

If you use an Apple Mac computer (i.e. iMac, MacBook Pro or MacBook Air), you can easily import your iPhone photos and keep them on your computer hard drive. When you plug your iPhone in via the USB cable, iPhoto should load with the iPhone's photo library present. Press 'Import' to add them to your iPhoto library.

Creating Photo Albums on Your iPhone

All the photos and videos you take are automatically stored in your Camera Roll but, after a while, this can become very cluttered; you can then create photo albums to group together certain events.

1. Select the Photos App, select albums and hit the + button in the top left corner.

2. Name the album and you'll be taken to the Camera Roll.

3. Tap the thumbnails to add blue ticks next to each photo and press Done when you have finished. A new album will appear in the Photos app.

4. In order to add additional photos to the new album, enter the Camera Roll, press Edit > Add To > Add To Existing Album.

Step 1: Album view shows your exisiting photo albums, and lets you add more.

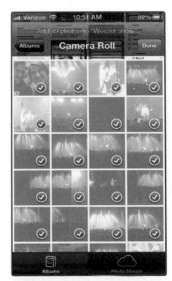

Step 3: Select photos from your Camera Roll to add to an album.

EDITING PHOTOS

Taking photos and recording video on the iPhone is just the start. They can be fine-tuned using a number of apps and then easily shared with friends and, indeed, the world at large.

Editing Pictures Using the Photos App

The Photos app features a few built-in editing tools that allow you to fine-tune your photos before sharing or printing them. Select a picture from your Camera Roll or an album and select Edit in the top right corner. The options that appear are the following:

Hot Tip

Rather than zooming in extremely close on a subject when taking the photo, keep the shot relatively wide and crop afterwards for the perfect proportions.

Above: Adjust proportions by cropping.

○ **Rotate:** Turns the photo in 90-degree increments. Great for ensuring that your pictures are the right way up.

○ **Auto-enhance:** A one-touch tool, which intelligently tweaks colour, contrast and brightness levels. This can be turned on and off.

○ **Red-eye Removal:** Hit this button and then begin prodding eyes to reduce red eye.

○ **Crop:** Drag the corners of the grid to trim out unwanted elements of the photo.

When you've completed the edits, press 'Save' in the top right corner.

More iPhone Editing Apps

As already mentioned, the built-in photo editing tools on the iPhone are very limited. However, there's a host of downloadable applications that offer much more flexibility when editing the look and feel of your photo before you share it. Here are a few available from the iTunes App Store.

○ **iPhoto for iPhone**: A mobile version of the Apple storage and editing app. You can adjust colours, add effects and filters, use Brushes to get rid of blemishes, and create attractive photo albums. £2.99.

○ **Adobe Photoshop Express**: A scaled-back version of the desktop software, this free app allows quick adjustments to colour, exposure, contrast, brightness, size, etc. All changes are saved back to the Photos app as a new photo.

○ **Instagram**: Primarily a photo-sharing social network (sign-up required), the free Instagram app is hugely popular and enables iPhone users to edit their photos using retro-style filters and effects (see photo sharing). It will also save uploaded photos back to your Camera Roll.

Above: Use the free Instagram app to edit, share and add affects to your photos.

Above: The Photo sharing screen presents you with options such as setting an image as your Wallpaper.

Sharing Photos

Throughout chapters two and three we've explained how to share photos, through Email, Messages, Facebook and Twitter, via the Share icon that appears whenever you're viewing a photo in the Photos app. There are a few more, however, that we have yet to encounter when you press the Share icon.

○ **Use as Wallpaper:** Bored of the Apple default wallpapers? You can easily set images from your Camera Roll (favourite place, dog, family, etc.) as your background. Scale the photo to your preferences and select Wallpaper, Lock Screen – or both.

○ **Print:** If your iPhone is configured to work with a Wi-Fi printer, you can instantly print off a snap from your Photos app.

○ **Assign to Contact:** Got a photo you'd like to see every time you get a call or email from a contact? Select this option and match it up with a Contact when the app pops up.

○ **Copy:** Hit this to paste the photo within another application, such as a third-part email app like Gmail that doesn't sit within the Share menu.

PHOTO STREAM

Photo Stream is the part of iCloud that is responsible for syncing and backing up your photos. If you enable Photo Stream (Settings > iCloud > Photo Stream), every picture you take using your iPhone will be automatically uploaded to iCloud and viewable across a host of devices, such as an iPad or the iPhoto app on your Mac if you have one.

Sharing Camera Roll Photos to Photo Stream

Once Photo Stream is enabled, you don't have to do anything to upload a photo. It's all done automatically, providing your iPhone is connected to Wi-Fi (if you're not, the iPhone will still upload the pictures next time you are).

Viewing the Photo Stream

Within the Photos app, there is a tab for Photo Stream. All of the photos you've shared to Photo Stream from your various Apple devices (not just those taken using the iPhone) are visible and sharable from this tab.

Viewing Photos from Other Apple Devices

Photo Stream – and indeed iCloud in general – is most useful to people who have multiple Apple devices. For example, if you have a Mac computer, photographs you import to the iPhoto program from your dedicated digital camera (Canon, Sony, etc.) can be automatically uploaded to Photo Stream (iPhoto > Preferences > Photo Stream > Automatic Upload) and will appear in your Photo Stream on the iPhone within seconds. This makes it easy for you to carry around the best photographs you've taken recently, where they're viewable and shareable. Best of all, the photos are stored in the Cloud so don't take up valuable space on your iPhone.

Above: Photo Stream backs up your photos on iCloud.

Shared Photo Streams

Although Photo Stream can be for personal use, iOS 6 users can set up Shared Photo Streams to share photos and albums with others. Friends who also have an iPhone or iPad will see the photos on their devices; alternatively, they can be viewed on the web at icloud.com. Here's how to set up shared Photo Streams.

1. Go to Settings > Photos & Camera and switch on Shared Photo Streams.

2. Go to the Photos app and select Camera Roll.

3. Hit the Edit button and select the photos you'd like to share.

4. Select the Share button and pick Photo Stream.

5. Choose a contact, name the album and select whether to make it available on a Public Website. This will allow anyone with a link to view it on icloud.com.

Step 4: Select photos from your Camera Roll to add to a shared Photo Stream which contacts can access from their own Apple devices via iCloud.

6. Select Next to load a new note with the attached picture. Add a comment, if you wish, and then press Post.

7. The new photo/album will appear in the Photo Stream section of your Photos app and in that of your recipient if they have an iOS device. They can then comment on and 'like' individual pictures.

8. For friends without iPhones or iPads, select the blue arrow next to the Shared Photo Stream and then press Send Link to share via Email, Messages, Facebook and Twitter.

Step 6: Add comments to images in your shared Photo Streams.

EDITING AND SHARING VIDEO

Videos shot on your iPhone's HD video camera can also be tweaked and sent out to the rest of the world – or just your mum.

Trim Video Recordings

The iPhone's built-in video editing options are less extensive, but the app does allow you to trim the beginning and end of a video. The Trim tool dispenses with the unwanted portions, making the clip nicer to look at and easier to share (due to the lower file size).

1. Select the video clip from the Camera Roll. There are markers at the beginning and end of each video. Start to drag the markers to where you'd like the clip to start and end.

2. The video timeline bar will turn yellow and a new Trim button will appear.

3. Once the clip is tailored to your satisfaction, press the Play button to preview the new edit.

Above: Edit video recordings using the Trim tool.

4. Once you're happy, press Trim. You'll be asked whether you'd like to Trim Original or Save As New Clip; the latter will keep both versions.

5. The new clip will then be ready to share.

Hot Tip

For a more precise edit, hold your finger down close to the Start–End indicators to zoom in on the video timeline.

Sharing Videos Via Email and Messages

The built-in options for sharing video are a little more limited than those for photos, mainly due to the increased file size. In order to share a video, select the clip from the Camera Roll and hit the Share icon. Select Email or Messages to load the video in a blank email or message, add the contact details and accompanying text, and press send.

Hot Tip

If you're on a limited mobile data plan, try to share videos via Wi-Fi.

Step 4: You can choose whether to publish a video to YouTube in Standard Definition or HD.

Publish Videos to YouTube

Within the Share options for videos in the Camera Roll you can also upload the video to YouTube, giving it a much larger potential audience. Here's how to do it.

1. Select a video from the Camera Roll and hit the Share Icon.

2. Select YouTube and type in your username and password. You can register for a free account at youtube.com, although Gmail users can use their Google login.

3. Type a Title and Description for the video in the respective fields.

4. Select Standard Definition or HD (Wi-Fi needed) quality.

5. Add tags (optional) to make it more visible within YouTube searches.

6. Make the video Public (available to anyone), Unlisted (available only to people with whom you share the link) or Private (only people you invite can view it).

7. Select Publish.

8. Once the publishing process is complete, you'll see a Banner notIfIcation. You can View on YouTube or Tell a Friend (which launches a new email, complete with the YouTube link).

Step 8: Use your iPhone to view your video on YouTube.

Sharing Video to Twitter and Facebook

Although the Photo app's Share Menu for videos does not feature the option to share to Facebook and Twitter, it's easy to do so via the official apps.

Above: You can share a video over Twitter by inserting it into a tweet.

○ **From Facebook:** Select the camera icon, select a video from the Camera Roll and press the new message icon in the bottom left corner.

○ **From Twitter:** Hit New Tweet, select the camera icon and choose either Take a photo or video or Choose from library. The former will take you straight to the Camera app and allow you to shoot a video to insert into your tweet.

Hot Tip

When uploading to social networks, use the trim tools to keep it concise, thus making the upload quicker and easier.

MUSIC

The iPhone itself rose from the success of Apple's iPod. Hands up if you've owned an iPod? OK, everyone put your hands down now. The iPhone brings together the great audio quality and usability of an iPod, a music store and more.

GETTING MUSIC ON YOUR iPHONE

With the iTunes Store, your own existing music collection, music streaming apps such as Spotify and radio services like iHeartRadio, there is literally no limit to the amount of aural satisfaction you can achieve using the iPhone.

Downloading iTunes

The iTunes software for PC or Mac is the iPhone user's best friend. It acts as a home for your digital music library and lets you easily transfer it on to the iPhone. If you've ever owned a digital music player – such as the iPod – the chances are you've got a stack of digital music files just waiting to be transferred on to your iPhone. Here is how you download iTunes.

1.   Go to itunes.com and select Download.

2.   On the next screen, add an email address, deselect the boxes (if you don't want junk email) and select Download Now.

3.   Depending on which web browser you're using, you'll see an indicator of the download's progress. When the download is complete, select Open or locate the iTunes file in your computer's Downloads folder.

5.   Follow the on-screen installation instructions.

Above: iTunes is a great digital library for your music, select albums and tracks to transfer to your iPhone.

Registering for an iTunes Account

In order to buy music, videos, apps, books or magazines from Apple's official portals, you'll need an Apple ID. We covered this briefly in chapter one, but here are the specific steps you'll need to take using your iPhone:

1. Open Settings > iTunes & App Stores.

2. At the foot of the Sign In screen you'll see a Create New Apple ID. You'll need to choose the region (e.g. UK) and hit Next.

Step 2: When setting up your Apple ID, you will be asked to select a region.

3. Add an email address and password, and choose three security questions and answers.

4. On the following screens you'll be asked to enter payment details and a billing address.

5. Apple will send you an email to verify the account. You'll need to click the link within the email and enter your username and password.

6. Once the account has been verified, return to Settings > iTunes & App Stores and enter your new account details.

7. Now you are ready to start downloading music.

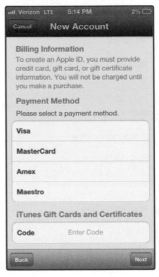

Step 4: You will be asked to enter payment details for your Apple ID.

Transferring Music from iTunes

Now that you have iTunes installed on your laptop or desktop computer, you can plug in your iPhone and begin transferring all of that lovely digital music on to the device – here's how.

1. Plug your iPhone into your computer using the bundled-in charging cable. iTunes should load automatically and your iPhone will appear as a button in the top navigation menu in iTunes 11.

Hot Tip

If you have a large library and only want to sync certain artists to your iPhone, you can choose Selected playlists, artists, albums and genres from the Sync music tab and apply each manually.

2. Click the iPhone and then hit Music in the top menu.

3. Select the Sync Music tick box. Pressing Sync to confirm will transfer your iTunes library on to the iPhone.

4. Next time you plug your iPhone into your computer, it will automatically sync your music to the phone, adding new songs and adjusting tweaked playlists.

Hot Tip

No digital music? Place your CD collection into your computer's disc drive. The CD will load in iTunes, thus allowing you to copy a digital version to your computer.

Above: Select 'Sync Music' to transfer your iTunes music library from your computer to your phone.

iTunes Match

Got too much music to fit on the iPhone? Apple has a solution for you, through iCloud. iTunes Match subscribers have access to their entire music library on the go, without having to store it on the phone. The service scans your hard drive for all digital copies, matches them up with high-quality files from its store and lets you stream them over the internet.

To subscribe for £24.99 a year, open iTunes on your computer, select iTunes Match from the vertical menu and use your Apple ID to sign up.

iCloud Music

iCloud users can instantly re-download previously purchased music from iTunes on to their iPhone. Open the iTunes app, select More and hit Purchased. Select the song/album and hit the

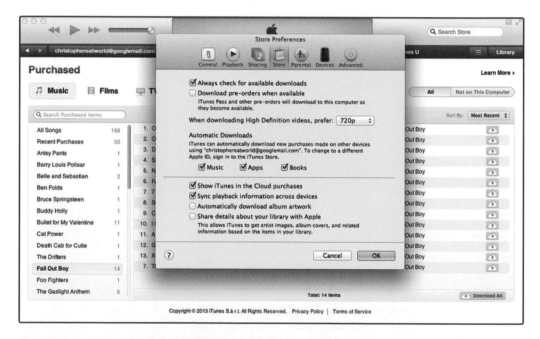

Above: Configure Automatic Downloads so that any items purchased on iTunes are downloaded straight to your iPhone.

Cloud icon to download. You can also configure iTunes to download automatically all your purchases to all devices associated with your Apple ID. For example, if you buy an album through iTunes on your PC, it can automatically be downloaded to your iPhone (and vice versa) without you having to lift a finger.

Here is how to configure Automatic Downloads.

1. Open iTunes on your computer and hit Purchased, under the Store heading in the left menu.

2. Select Configure Automatic Downloads.

3. Tick Music from the list.

THE iTUNES MUSIC STORE

Thanks to the iTunes app that lives on your iPhone, you're able to download, play and own virtually any noteworthy song or album ever released. From the contemporary to the classic, it's all there waiting for you under one roof.

Buying Songs and Albums from iTunes

Hit the purple iTunes icon on your Home screen (it's not the orange music app!) and, if it's the first time you've accessed the app, it will open with a splash page showcasing the store's new releases and featured artists.

Above: View albums on iTunes to preview and purchase songs.

1. In order to search for specific songs, albums or artists, you can hit the Search tab at the foot of the screen.

2. Type the keywords of your choice (e.g. Mozart or Metallica) and hit Search.

3. The search will yield results applicable to your keywords and you'll see a list of songs and albums. Next to each item, you'll see a button displaying the price. Hit this and select Buy.

5. Once you've entered your Apple ID password, the item will begin downloading and it will appear within the Music app (see page 187) when complete.

Previewing Songs

The iTunes app allows you to try before you buy, by offering 90-second previews for songs. Just touch the name of a song and the preview will load and begin playing.

Chart Music

As well as searching for your favourite music, the iTunes Music Store also allows you to pick up the songs and albums currently topping the charts. Load the iTunes app and select Charts from the top menu to see the top songs, albums and music videos. Follow the instructions on the previous page to buy this music.

THE MUSIC APP

Now you have added your favourite music to the iPhone – either through transferring or buying from iTunes – you're probably ready to start playing it. The iPhone's Music app (called iPod on older models) plays home to your new library. Open it and let's get started!

Choosing Music

Once you open the Music app, there are several different ways of accessing your music. The simplest option is probably the Artists tab at the foot of the device, which lists bands in alphabetical order.

1. Click an artist's name from the list.

2. The next screen will display the albums that you own.

3. You can pick an album or select All Songs to begin playback.

In order to begin playing all songs on your phone in a random order, select Songs and hit Shuffle.

Above: Select an artist's name to view and play their albums and songs.

The Now Playing Screen

When you select a song, album or playlist, what is currently playing will appear in a new screen, allowing you to control playback, volume, the next song and more. Here's a rundown of the media controls from within the Now Playing screen.

- **Cover art**: The album cover.

- **Play/Pause**: Hit this icon to pause and restart playback.

- **Previous/Next**: Move to the previous or next track on the list (hold these buttons to rewind or forward through an individual song).

Above: To browse your music using CoverFlow, flick one way or the other on the screen; flick fast to move through more covers, flick more slowly to move one at a time.

- **Volume**: The bottom slider allows you to control the volume of a track. This will move up/down when you use the physical volume keys on the side of the device.

- **Repeat**: Selecting the Repeat icon (the arrows eating each other's tails) once will repeat the current album, once it has finished playing. Hitting it twice will repeat the current track until you either die or get bored.

- **Shuffle**: Hit this icon (two arrows intersecting with each other) to shuffle the current song queue (playlist, album, etc.).

- **Song progress**: This slider shows you how much of the song has played and how much is left. You can move this to skip to the awesome guitar solo!

- **Title bar**: Features the artist, song title and album.

- **List view**: Select the list icon in the top right corner to change the view and see the entire album or playlist.

- **Back arrow**: Return to the previous screen.

Multitasking With Music

Once you've selected your music, you're free to leave the app and send emails, browse the web or use most other apps without interrupting playback. You can return to the Music app to stop or change music but there are a couple of handy shortcuts to assist with this.

- **While the phone is unlocked**: Double-click the Home button to launch the Multitasking Bar and slide it to the left to reveal a music control screen. As you can see to the right, you can pause or switch songs, or select the Music icon to return to the app.

- **While the phone is locked**: Double-click the Home button. You'll see the cover art from the item you're currently playing and another set of media controls.

Creating Playlists

If you've synced your iTunes library by plugging your phone into your computer (see page 182), all of the playlists you've created in iTunes will be carried over to the iPhone too. You'll see them in the Playlists section of the Music app. However, from within the Playlists section of the Music app, you can also add new ones.

> **Hot Tip**
>
> **Within the music app, at any time, turn the iPhone on its side to launch CoverFlow view, allowing you to flick through album covers and select them. Turning the phone back upright will revert to the previous view.**

Above: The Multitasking bar displays music controls at the bottom of the screen, allowing you to leave iTunes.

Step 1: Enter a name and touch save, you can then begin adding songs to your new playlist.

1. Select Add Playlist, give it a name and press Save.

2. The Songs section will load. Press the blue + button next to a song to add it to the list.

3. To add all songs from an artist or album, select them and hit Add All Songs.

4. Press Done when the playlist is complete.

5. Next time you plug your phone into iTunes, the Playlist will sync back to iTunes too.

OTHER MUSIC SERVICES

Although iTunes is a one-stop shop for your music transferring, purchasing, streaming and listening needs, there's a world of fantastic music services out there. Here are some favourites.

- **Spotify**: offers unlimited access to 20 million songs (monthly subscription required).

- **BBC iPlayer Radio**: live and on-demand access to the BBC's vast library of radio stations.

- **Shazam**: tag a song you don't know and it will tell you the title and artist (free from the App Store).

Above: Listen to your Spotify playlists from your iPhone.

○ **Podcasts**: free, self-contained radio shows, available to download and play on your iPhone through Apple's Podcasts app.

SHARING MUSIC TO SPEAKERS

If you're at home, you can play music through the iPhone's speakers; they're actually not bad on the iPhone 5 and can reach a decent decibel level. If you're in public you can use any set of headphones with a 3.5 mm jack. However, if you're looking to crank up the volume, you can broadcast your music super loud with the aid of a speaker.

Speaker Docks

Made popular in the iPod era, speaker docks allow you to place your iPhone into a connector in order to play music through a more powerful set of speakers. Speaker docks are great, because they'll also charge your iPhone while it is docked.

AirPlay Speakers

AirPlay is a wireless technology developed by Apple to share music and video over Wi-Fi networks. If you have

Hot Tip

iPhone 5 owners be careful! Due to the new Lightning connector on your handset, you'll need a 30-pin-to-Lightning adaptor to connect to older accessories.

Left: Some AirPlay speakers allow you to listen to music and charge your phone at the same time.

AirPlay-enabled speakers (they're available from Bose, B&W, Sony, JBL, Phillips and more), you can send any music from your iPhone to the speakers without plugging in.

If an AirPlay speaker is connected to the same Wi-Fi network as your iPhone, you'll see the AirPlay icon in the Now Playing screen. Click this to select the name of the speaker.

Hot Tip

If you own an Apple TV set-top box, you can send music directly to your television set.

Bluetooth Speakers

You can also use the iPhone's Bluetooth connectivity to connect to speakers – here's how.

1. Make sure that Bluetooth is enabled on the device you'd like to pair with.

2. Enter the iPhone Settings and select Bluetooth. Toggle the switch to turn Bluetooth On.

3. The iPhone will go into Discovery mode search for any speakers with Bluetooth connectivity.

4. When you find your speakers (usually under the manufacturer's name and model number), select them.

5. You may be asked for a four-digit PIN code to connect to the speaker (it'll be in your instruction manual). Once the initial connection is made, automatic connections will be made in future.

6. Enter the Music app (or Spotify or internet radio apps, etc.) and start playing. Audio should now be coming through the speaker.

Wired Speakers

A more primitive way to connect to speakers is through the iPhone's headphone jack. A pair of PC speakers, for example, should have a cable that will plug right into the iPhone.

VIDEO

The iPhone is a staggeringly powerful tool for video. Not only can it record video in full high definition, but you can also play your favourite TV shows and movies, stream them from the internet and purchase brand new films to rent and own.

ADDING VIDEO CONTENT TO YOUR iPHONE

If you read the section in this chapter on syncing your music files, then you'll soon notice that the process of getting video content on to your phone is pretty similar. It can be bought or streamed directly from the phone, or added from your existing library.

Syncing Video Via iTunes

Just as we did with music, it's possible to transfer existing videos you may have previously purchased or added to iTunes directly on to your iPhone.

1. Plug your iPhone into your computer using the USB cable; iTunes should automatically load (if you don't have iTunes, *see* page 180).

2. Select the iPhone from the top navigation menu.

3. From the iTunes screen select Movies or TV Shows. You'll see a list of videos you currently have within your iTunes library.

Hot Tip
In order to add video content to your iTunes library from your computer, open iTunes, select File > Add to library. To work, the video will have to be in the formats described on p. 196.

4. Tick Sync Movies/TV Shows and then begin selecting the content you'd like to add to your iPhone.

5. Alternatively, tick the Automatically Include box and select from the drop-down menu to set rules (all movies, most recent, unwatched, etc).

6. The Capacity bar at the bottom of the iTunes screen will tell you how much space you have on the device and will update as you add video files.

7. When you're done, press Sync and the movies or TV shows will begin to load on to your device. Don't unplug the iPhone until this process is complete.

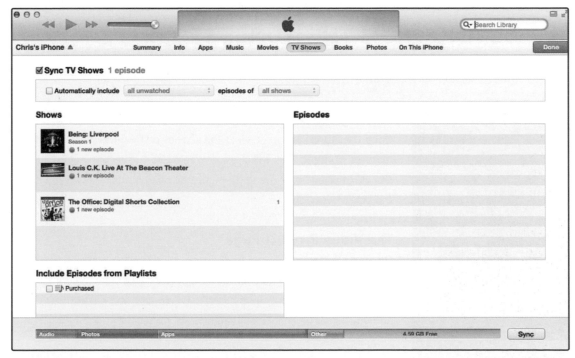

Above: Sync your TV shows over iTunes so that you can watch them from your iPhone.

iPhone/iTunes Video Formats

Apple likes to keep things tidy, but it also likes you to buy things from them. To that end, it limits the video formats that work in iTunes/on the iPhone. The files need to be .MPEG-4 or .H.264 files to play on the device. Therefore, videos in other popular formats, such as .WMV, .AVI and .MOV, will not work on the iPhone or in iTunes. However, they can be converted using software such as Handbreak (www.handbreak.com).

BUYING FROM THE iTUNES STORE

As is the case with music, you can purchase a world of movie and TV content through the iTunes app on your iPhone.

Finding Video Content

Open the iTunes app and select the Movies or TV Shows tab to see featured content. This will show new releases and recently aired shows. In order to locate content, hit the Search button and type in the title of your choice.

The Video Product Page

Once you've found the movie of your choice, you'll be taken to its product page where you'll have the opportunity to view trailers. You'll also see a plot summary, cast and crew list, and a certificate, and you'll get the chance to read reviews from other users.

Left: Video product page gives added info about movies or TV shows featured on iTunes.

Buying to Own

Buying video content from the iTunes app means that the digital file is yours to keep forever. You can transfer it to as many Apple devices as you please and it will always be associated with your Apple ID if you want to re-download it.

1. Identify the movie or TV show you wish to own. Click the Buy button and enter your account password. Once that's entered, the file will start downloading.

2. You can see its progress by selecting More > Downloaded within iTunes and it'll give you an indication of how long is left to download.

3. In order to start watching as the movie is downloading, exit iTunes and enter the Video app. It'll be waiting for you.

4. Select the title to begin playback.

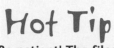

Hot Tip

Be patient! The files are usually well over 3 GB in size so they will take a while to download completely. Movies can only be downloaded over Wi-Fi.

Renting Movies

iTunes movie rentals are perfect if you just want to watch a movie once rather than have it to own. The price is around the same as renting a DVD from your local video store. In iTunes, you follow the same procedure as buying a movie (unfortunately, TV shows can only be bought to own) except that you hit the Rent button from the product page.

There are a couple of caveats with renting movies, though. Once you begin to watch the film, it has to be finished within 24 hours. Otherwise, you have 30 days before the rental expires. The Video app will tell you how long you have left to view the video.

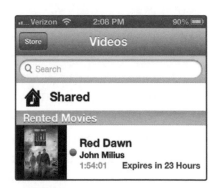

Above: The Video app tells you how long you have left to view a rented iTunes movie.

HD or SD?

Prices for movies and TV shows are automatically displayed for the HD version but if you're not concerned about the highest quality, you get the video cheaper by scrolling to the bottom of the product page screen and selecting Also Available in SD.

Above: For TV series you can either buy a pass to view the whole season. or scroll down to select a certain episode.

Season Pass or Episodes?

If you're buying TV shows from iTunes, you'll have the option to pick up the latest episode or to buy a pass, which gives you the entire season. The first option is great if you have a favourite episode or you missed one when it was aired on TV. The season homepage will point you towards buying the whole season but simply scroll down the page to access particular episodes.

iCloud Movies

Since movie and TV show files are so large (just one movie could take up a huge section of your iPhone's available storage), you won't be able to fit multiple films on the iPhone at one time – especially if you've only got the 16 GB model.

Above: You can download previous movie purchases to your phone by connecting to Wi-Fi and using iCloud.

As we explained earlier, it's easy to add previously purchased video files using the iTunes computer program. However, iTunes in the iCloud keeps a record of everything you've bought from the store (either from the iTunes computer app for Mac and PC on your iPhone or iPad) and lets you re-acquire the content wirelessly. In order to re-download any purchases you've made, go to iTunes > More > Purchased and select the iCloud download button from the title page. The file will then be downloaded to your device.

THE VIDEO APP

The iPhone's Video app houses all of the content you currently have stored on the device, either transferred from iTunes or bought/rented from iTunes, as explained in the previous section. Everything appears in list format, under Movies, Rented Movies, TV Shows and Music Videos.

Playing Movies

The video playback screen on the iPhone is a lot simpler than the music equivalent. Selecting a video from the app will instantly start playback and bring up the following controls.

- **Play/Pause**: Stop and restart a video.

- **Rewind/Forward**: Press the rewind button once to restart the video or hold it down to scan backwards. Hold the Forward button to move forwards.

- **Volume slider**: Drag left or right to alter the volume (you can use the volume keys too).

Above: Playing movies in the Video app is simple. When you touch the screen it will display play, pause, rewind and forward buttons, along with a progress bar.

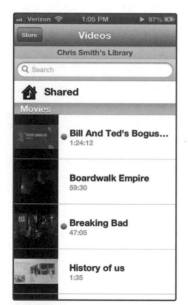

Above: Your shared library contains items from devices which are connected to the same mobile network.

- **Progress meter**: Indicates how far into the video you've progressed. Grab the circle and drag to move backwards and forwards.

- **Done**: Hit this to return to the Video app.

Hot Tip

Playing video can be a big drain on battery life. To ensure you can watch for longer, turn down the screen brightness and close down apps that aren't in use.

iTunes Sharing over Wi-Fi

If your iPhone and computer are connected to the same mobile network, you can stream items from your iTunes library directly to your iPhone. Open the Video app and select Shared.

Select your Library and all of the video files listed on your computer will show up; you can then play them as normal.

AirPlay Via Apple TV

When you're watching a movie on the train on the way home from work, wouldn't it be nice to just send it to your television when you get through the door and relax in your favourite chair to watch the ending? If you have an Apple TV set-top box, that's exactly what you can do.

We mentioned AirPlay in the Music section but it extends to video for Apple TV owners too. Videos from most apps and an increasing number of websites now have AirPlay support. Here's how to share video with your TV.

- You'll need to be registered on the same Wi-Fi network as the Apple TV.

- If it's in range, you'll see the AirPlay icon in the video playback controls.

- Hit this and select Apple TV.

- The video will automatically be transferred at exactly the same point in your viewing.

Left: Selecting Apple TV allows you to use AirPlay to transfer content to your Apple TV at home.

VIDEO STREAMING SERVICES

Although Apple would love you to rent and buy videos only from the iTunes Store, there are plenty of other alternatives on the App Store for streaming video over the internet. Here are a some of our favourites.

Above: The LOVEFiLM app allows you to create priority lists for movies and TV.

○ **LOVEFiLM**: access thousands of movies and TV shows through the free iPhone app, available from the App Store. Subscriptions start at £4.99 a month (www.lovefilm.com).

○ **Netflix**: watch movies and TV on your iPhone for £5.99 a month (netflix.com).

○ **YouTube**: there's a dedicated YouTube iPhone app available from the App Store, featuring billions of user-uploaded videos.

○ **BBC iPlayer**: the iPlayer app allows users to get free access to most BBC TV shows from the last seven days. Programming can be downloaded over Wi-Fi and watched off-line.

○ **Sky Go**: this allows live streaming of a host of channels (including Sky Sports, Sky Movies and Sky One) over Wi-Fi and 3G (you need to subscribe to Sky TV).

Above: The YouTube app is a great way to browse and watch recently uploaded videos.

READING

If you have an iPhone, you'll never be short of something to read – that's a promise. From brand-new bestsellers to the classic books of yesteryear, the iBooks store is your oyster, while Newstand provides instant access to many popular magazines.

BOOKS

We've come a long way since the invention of the printing press. Now a large proportion of the books available at your local bookstore can simply be downloaded to digital devices such as the iPhone and Amazon Kindle e-reader. Over the next few pages, we'll explain how to load new titles on to your smartphone.

THE APPLE iBOOKS STORE

Apple's digital bookstore, built into your iPhone, is called iBooks. Selecting this app will load a bookshelf which you'll find is currently empty. In order to rectify that, hit the Store button in the top right corner of the app.

Finding Books on iBooks

When you load the iBooks Store for the first time, you'll see a splash page similar to that you may have seen when accessing music, movies and TV shows. It showcases featured content, new releases and upcoming titles for pre-order. However, there are several other tabs at the foot of the store.

Left: Search the iBooks store for free and paid books available for instant download.

Above: You can search for a book by typing an author's name or book title into the search bar.

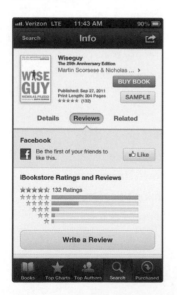

○ **Top Charts:** Hitting this tab will automatically load the store's most popular paid and free books from all categories. You can swipe right on either list to see the top 100 or select See All at any time to display the titles in a list.

○ **Top Authors:** Hitting this tab will load a list of authors, which looks similar to your Contacts book. Select an author from the list to load their titles.

○ **Search:** The easiest way to find a specific book is through the Search tab. Simply type the name of the book or the author to load the relevant titles.

Hot Tip

Hit the Categories button to narrow down the charts based on genres (biography, romance, etc.)

Buying a Book from iBooks

Once you've identified the book of your choosing, you can buy it from the product page. The iBooks product page is very similar to music and video store pages, as it features tabs for descriptions, reviews and related titles. Hit the Buy Book button, enter your Apple ID password and the book will be downloaded, usually within a minute. It will then appear in your bookshelf.

Left: When you select a book you will be presented with the green Buy Book option. Touch this button and enter your Apple ID to pay for the book and download it to your bookshelf.

Book Samples

When browsing in old-fashioned brick and mortar bookstores, it's only natural to pick up a book and flick through a few pages before handing over your cash. The iBook store has an equivalent through the Sample feature.

On the product page, you can hit Sample and a few pages will be downloaded to your virtual bookshelf. Many books even let you read the entire first chapter before committing to buying.

Free Books

Publishing laws mean that once a published book has been around for a certain amount of time, copyright expires and it becomes freely available. The free books section of the iBooks store features plenty of junk, but there are also many gems. You can download free books from the likes of Charles Dickens, Jane Austen, Mark Twain, William Shakespeare and more.

iCloud and iBooks

Like most of the multimedia elements we've encountered in this chapter so far, Apple's iCloud solution means that all previous purchases are available to read on your new iPhone too. The Purchased tab lists the books that you may have bought through iTunes on your computer or through the iBooks app on the iPad. Hit the cloud icon to the right of the book to add these titles to your iPhone's bookshelf.

Above: There are many free books available to download to your shelf.

READING iBOOKS

Now you have stocked your virtual bookshelf with a host of titles from the iBooks Store, it's time to start reading. Select the thumbnail cover from the Library to load the book full-screen on your device.

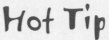

Hot Tip

If you ever get bored of the page turning mechanism (we haven't), you can just tap the screen to turn a page.

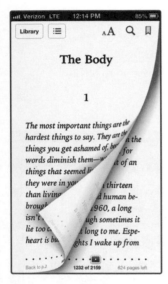

Above: You can turn the pages of your iBook by dragging a page to the left.

Turning Pages

The pages in iBooks turn as if you were reading a real book and it's one of the best-looking things you can do on an iPhone. Use your thumb to slowly drag from various points on the right side of the page to see this beautifully imagined feature in all its glory. You can also give the screen a little flick when you're ready to turn a page. Naturally, when going back a page, flick from left to right.

Finding a Page

If you're searching for a particular page within a book, there are a number of ways to reach your destination quickly.

○ **Contents:** Hit the list icon at the top of the screen, where you can access the Contents page. Tap an item from the page to head to the beginning of that chapter.

○ **Scan:** At the foot of the screen you'll see a horizontal bar. Drag this back and forward to find specific pages. Take your finger off the screen when you've reached your destination.

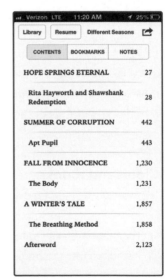

Above: Touch the list icon at the top of your screen to display the book's contents page.

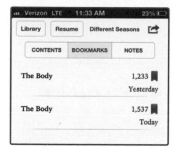

Above: View your bookmarks.

○ **Search**: Hit the Search icon at the top of the screen and type in words or a page number. Make your selection from the list of results.

○ **Bookmarks**: You can add as many Bookmarks as you wish to a book, just by pressing the icon at the top of the screen. All Bookmarks can be accessed by hitting the list icon and selecting the Bookmarks tab.

Changing the View

iBooks enables you to tailor your reading experience depending on your environment, eyesight and personal preferences. Hit the small and large 'A' icon to load the options, as follows:

○ **Brightness**: Rather than exiting the app to change the brightness, you can do it directly from the iBooks app. This is useful when you change your environment from bright to dark.

○ **Text size**: Hit the 'A' buttons to increase or decrease the size of text to make it more readable.

○ **Fonts**: iBooks boasts seven fonts. Pick your favourite from the list.

○ **Themes**: Here you can choose the colour schemes. The options are: black text on a white background, brown text on a sepia background or light grey text on a black background. The latter will allow you to read at night without burning out your retinas.

Above: Touch the A icon to alter viewing options such as brightness and theme.

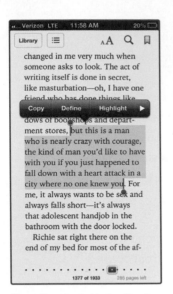

Above: You can select text sections to copy, define or highlight.

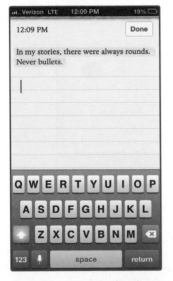

Above: You can annotate selected text sections in a separate window.

- **Book or Scroll**: From the Themes menu you can also choose the Book view (which requires you to turn pages) or Scroll, which presents all text from the book on one continuous page.

Adding Highlights, Copying Text and Making Notes

iBooks also allows users to annotate the text with notes and highlights, while sections can also be copied and shared with friends or on social networks. In order to use these features, you'll need to select a piece of text. Hold the screen and move the two blue markers to highlight the relevant section of text. You'll see an options tab pop up, allowing you to do the following.

- **Copy**: This will copy a segment of the text, allowing you to paste it elsewhere in another app (email, document, social network, message, etc.).

- **Define**: If you've highlighted a particular word, you can ask iBooks for a dictionary definition.

- **Highlight**: Select this button to highlight a section of text. This is a great feature if you're reading for educational purposes. Hitting Highlight will turn the section yellow but it will also bring up a new options screen, allowing you to change the colour, delete the highlight, add a note or share it.

- **Note**: The first option after pressing the 'next' arrow allows you to annotate text. The selected text will open in a new window. Type your notes and press Done to return to the page. In order to edit what you have written, hit the post-it note next to the section.

- **Search**: Hitting 'Search' will give you the option of searching for the passage elsewhere on your phone, or via the web (Google) or Wikipedia.

- **Share**: Share a favourite passage via the usual means.

> ## Hot Tip
> **All Notes and Highlights will appear within the Notes tab next to Contents and Bookmarks.**

KINDLE

Many people who graduate to using an iPhone own or have owned an Amazon Kindle ebook reader and are likely to have a library of digital books they've already bought. Thankfully, there's a Kindle app for the iPhone where users can access all of their previous purchases.

The Kindle App

This application can be downloaded for free from the App Store (see page 155). Once you have opened it, enter your Amazon username and password, and select 'Register this Kindle'. All of your previous purchases will be listed so tap the cover to download each item. Unfortunately, you can't buy books from the Kindle app; instead, you'll need to do that from amazon.com. Fret not, though: after purchase, the books will still appear in the iPhone app.

Reading a Kindle Book

Kindle Books can take advantage of Amazon's neat Whispersync technology, which means whenever you open a book in the Kindle app, it will sync to the last page you read on any of your devices. Beyond that, the means of reading books doesn't differ much from iBooks.

Above: Enter your Amazon username and password and tap Register this Kindle to begin using the free app.

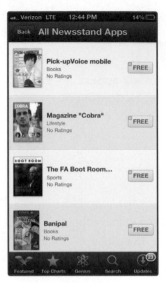

Above: Browse the Newsstand list to look for publications to download.

MAGAZINES

Beyond books, you can also have many high-profile periodicals delivered straight to your iPhone. This is done through the Newsstand app on the iPhone's Home screen.

Newsstand

If you've used the iTunes Store for buying music and movies, Newsstand will feel very familiar. Hitting the Store button within the app will take you to the featured content. You can browse New & Noteworthy and the All Newsstand Apps sections or use the Search tab to look for something specific. Once you've found a title you like, hit the thumbnail image to download it to your Newsstand.

Downloading Issues and Subscribing

Adding a magazine to your Newsstand app won't give you immediate access to any content – just the opportunity to subscribe for a particular period of time or download individual issues from within the app. Select the option of your choosing by clicking the price. All purchases will be charged back to your iTunes account, just like music and movies.

Above: You will be presented with the option of buying either a single issue or a subscription to items downloaded to your Newsstand.

Hot Tip
When a new issue is available to download, you'll receive an Alert notification from Newsstand and a number will appear next to the cover.

NEWSPAPERS

As with magazines, the leading UK newspapers are yet to embrace Newsstand on the iPhone, because it's difficult to format a full newspaper on such a small screen. However, all British dailies have dedicated apps available from the App Store. Some offer free access, whereas others will make you pay to subscribe.

Free Newspaper Apps

The *Guardian*, *Daily Mirror*, *Daily Mail*, *Daily Express*, the *Independent* and *Daily Star* all offer free apps that you can download from the App Store. These are, essentially, mobile-optimized versions of their websites which can also be accessed over the web through Safari.

Subscription Newspaper Apps

Some newspapers make you pay to access their content through the dedicated iPhone app. *The Daily Telegraph* (£9.99 a month), *The Times* (£9.99 a month) and *The Sun* (69p for the app download and then 69p per month) all have a paywall. Users can subscribe to the newspaper from within the app, which will be charged to their Apple ID and will auto-renew each month.

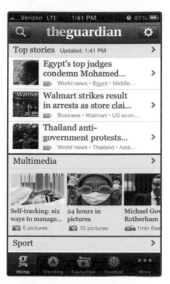

Above: The Guardian app is a free newspaper app designed for mobiles.

> ## Hot Tip
> To stop auto-renewals, open iTunes on your computer, select your account page and scroll to Subscriptions. Then hit Manage.

Other News Applications

Beyond the traditional daily newspapers, there are other ways to obtain your news through stand-alone apps on the iPhone. The BBC News app offers a brilliant design with written, audio and video content, while Sky News and Channel 4 News bring great video content from their broadcasts. All three can be downloaded from the App Store.

Above: The Channel 4 app shows video content from news broadcasts.

GAMING

The iPhone has given birth to a whole new era of hand-held gaming. Who needs a gamepad when you have a touchscreen? Within this section we'll help you to get the latest titles on your iPhone and show you how to become an Angry Birds master.

DOWNLOADING GAMES

All iPhone games are available to download from within the App Store (they don't have a dedicated store) and can be obtained using the methods explained in the previous chapter on apps.

PLAYING GAMES ON A TOUCHSCREEN

The multi-touch screen on the iPhone makes it a powerful gaming device. It means you can perform a host of swiping, tapping, zooming, pinching, dragging and flicking gestures using more than one digit at a time. The entire screen is your control pad and it helps to give each game its own unique flavour, while breathing new life into classic titles from yesteryear.

Above: Angry Birds is a very popular and addictive touchscreen game available for download.

We'll use a few of the App Store's most popular games as examples to help you get started and to explain the different ways to control games using the phone.

Line-up and Launch: Angry Birds

Angry Birds is the most popular touchscreen game ever but also one of the simplest to play and the hardest to put down. The idea is to use a slingshot to catapult birds in order to destroy the pigs that have done a runner with their precious eggs. Once you've downloaded the game from the App Store (there are free and paid versions), it's easy to get started.

Pull back the slingshot to aim the Angry Birds at the pigs and then release it to fire. You'll need to get the angle and the power right to hit your target; experiment and you'll get the hang of it within minutes.

Swiping and Slashing: Fruit Ninja

You play the role of a ninja who has a serious aversion to fruit. Use your thumb or finger as a sword to slash the falling fruits in two. The more fruits you can eliminate with one swipe, the higher your score. Once you get the hang of Fruit Ninja, try Infinity Blade where you need to use a sword against a host of beasties from another world.

Drag and Drop: Scrabble

Game developers have also reinvented classic board games, such as Monopoly, Battleships, Trivial Pursuit and even chess, for the iPhone. However, one of our favourites is Scrabble. You simply pick a letter and drag-and-drop it on to the board to form words, and the game will do the rest. You can play against the iPhone, against an opponent on the same phone or over the internet against a random opponent.

Virtual Buttons: FIFA Soccer

A lot of the games available on the Xbox 360 and PlayStation 3 consoles are also available for the iPhone. These scaled down versions feature virtual buttons and directional arrows, which allow you to pass, shoot, tackle and move players, in lieu of the traditional gaming pad.

Drawing and Painting: Draw Something

This Pictionary-inspired game was a mini phenomenon in 2012. The app provides you with a word and you use the touchscreen to draw a clue

Above: Scrabble has been redesigned for the iPhone. Play by dragging and dropping letters.

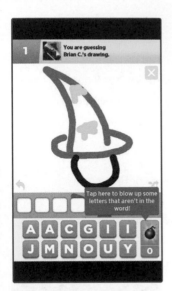

Above: The Draw Something game allows you to compete with others over the internet.

by using different brush tools and colours. When your masterpiece is complete, you send it to an opponent, via the power of the internet, and they have to guess what you've drawn.

Tilting and Turning: Temple Run

Addictive, Indiana Jones-inspired game, Temple Run employs a completely different control method to those previously explained. Thanks to the iPhone's gyroscopic sensor, you control the main character by tilting the phone left and right, as well as swiping the screen to jump, duck and turn. Many car racing games, such as Need for Speed and Asphalt, also use the tilting control method.

Hot Tip

Many paid games from the App Store have an ad-funded free or lite version, which enables you to try before you buy.

GAME CENTER

The Game Center app on your iPhone's Home screen allows you to challenge friends to games online, while it also keeps track of your personal accomplishments on each game you play. When you first load the app, you'll need to sign in with your Apple ID and password.

Achievements and Leaderboards

Each game you play on your iOS device will be listed in Game Center. If applicable, you'll see your worldwide ranking (sometimes it can be quite depressing reading!), global leaderboards and the achievements you've unlocked within each game.

Left: The Game Center app displays your games and leaderboards.

Hot Tip

When you first log into Game Center, allow the app to search your Facebook friends for those already using the app.

Adding Friends on Game Center

Once you've added your iPhone-toting friends to your Game Center account, you can view the games they're playing and challenge them to multiplayer battles. Alternatively, you can hit the + button to Add Friends individually from your contacts book.

Above: Game Center can search your Facebook friends to connect you with others using the app.

Challenge Someone to a Game

A new feature has been introduced in iOS 6 by Apple: users can challenge their pals or random users to beat their own scores in Game Center.

Multiplayer Games

While games such as Angry Birds and Temple Run are single-player affairs, a lot of games allow for multiple participants. Some games, like Scrabble, have their own mechanisms for finding online opponents; Game Center also helps to pair you up with partners or adversaries over Wi-Fi.

Select Multiplayer and Game Center from the game's screen, and Game Center will look for fellow participants. In some cases it can take a few minutes before players can be found.

Left: Selecting the Multiplayer option prompts Game Center to search for other players.

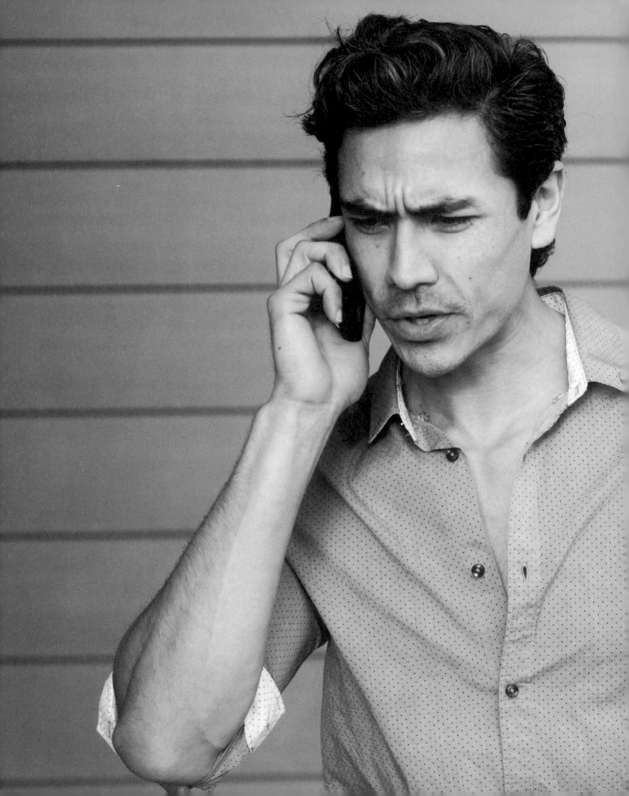

ADVANCED iPHONE

CUSTOMIZING

The ability to personalize your handset, customizing how it looks and behaves, is one of the biggest draws of the iPhone. Like a made-to-measure suit, you can refine almost everything the iPhone does so that it feels like it's designed to accommodate all your habits and idiosyncrasies.

MAKING IT YOUR OWN

Hit the Settings icon on your Home screen and you'll see hundreds of ways to make your phone behave the way you want. From changing the Lock screen wallpaper to which web browser you prefer, you can mould your iPhone to meet your needs.

Getting Started With Settings

To customize tools and apps, choose Settings from your Home screen. Here you'll see a list of everything you can amend, from WiFi and Sounds, to Privacy, Apps, Camera *et al*. Just click on the small > next to each item to dig deeper into its settings.

Above: The Settings page allows you to amend and customize many features.

PREFERENCES

Airplane Mode

Using your mobile to make calls in-flight is not allowed, but your iPhone isn't just a phone. Once you hit cruising altitude, you'll want to listen to music, watch a film or play a game and, luckily, you can. In Settings, just toggle Airplane Mode on before you take off. This shuts down the phone's wireless functions including Wi-Fi, 3G, Bluetooth and 4G. A small aeroplane icon will appear in the Status Bar to show you it's activated.

Managing Your Bluetooth Preferences

Bluetooth lets your iPhone communicate wirelessly with over devices such as headsets, speakers or phones. In order to manage your Bluetooth setup, go to Settings and select Bluetooth. From here you can toggle Bluetooth On and Off, and also see a list of devices. When Bluetooth is on, a small symbol appears in the Status Bar. If the Bluetooth icon is blue or white, the iPhone is communicating with a paired device, whereas if it's grey, Bluetooth is on but not sending data to a paired device.

Managing which devices your iPhone can communicate with is simple – just follow the steps below.

Above: Turning your Bluetooth on leads a small icon to appear in the settings bar.

- **To connect or disconnect a device:** Tap the name until the words Connected or Not Connected appear.

- **To unpair a device:** Tap the blue >, select Forget This Device and it will no longer appear in your list of available devices.

Going Private: Managing Your Privacy Preferences

Your iPhone uses two methods to pinpoint where you are: built-in sat nav-style GPS (Global Positioning System) or by triangulating signals from mobile phone masts and Wi-Fi points. These slightly sci-fi capabilities are used for everything, from helping you to navigate while using the Maps app to tagging photos you've taken with a location.

However, if you feel that this is all a bit more Big Brother than you're happy with, you can go 'off the grid'. Individual apps should ask your permission to use Location Services via pop-up messages but you can also turn off Location Services completely by going to Settings > Privacy and toggling to Off.

More Location Preferences

You'll find a System Services button at the bottom of the Location Settings screen. Tap this and you'll be given the option to turn off a bunch of settings that relate to your phone network, location-based iAds and time zone.

Notifications Settings

iPhone Notifications were made available with the launch of iOS 5. They provide a variety of different ways to alert you when something is received, updated or you've set a reminder. In addition to showing up as Alerts, Banners and Sounds, these updates also appear in the

Notifications Centre, which is accessed by swiping down from the top of the Home screen. You can fine-tune how and when your alerts appear in Notifications Settings.

○ **To assign Notifications Settings for an app**: Go to Settings > Notifications and select the app from the list. You can then toggle the Sounds, Alerts, Banners and Badges On or Off for that app, as well as deciding whether the alerts appear in the main Notification Centre.

Above: Notification settings allow you to alter alerts for individual apps.

Above: Sound settings allow you to customize all audio aspects of your iPhone.

Sounds Settings

Go to Settings > Sounds. From here you can switch audio alerts on or off, for everything from incoming email through to calendar alerts. You can also assign each function its own tone, manage your ringtones, set volume levels and decide whether your phone should vibrate when you have an alert.

Screen Settings: Adjusting Brightness

Remember this: the brighter your screen, the quicker your battery dies. Select Settings > Brightness & Wallpaper and use the slider to decide which you're most willing to sacrifice: a superb crisp bright screen or a phone that lasts longer.

Hot Tip
Want to make your iPhone battery last longer? Switch on the Auto-Brightness function in Settings > Brightness & Wallpaper and the iPhone will adjust itself depending on light conditions.

Wallpapers

Just like desktop screensavers, adding an iPhone wallpaper to your Home screen or Lock screen is a great way to personalize your handset. It also helps you to spot it on a table full of similar phones. The iPhone comes pre-loaded with lots of 'pretty' wallpapers but you can also use pictures you've shot or downloaded to your photo albums. In order to assign wallpapers, go to Settings and choose Brightness & Wallpapers then select either the Lock or Home screen. Choose your image and press Set.

About Your Phone

Want to know how many songs, videos, photos and apps are on your phone? Or how much space you have left to store more content? You need to check out the 'About' section in Settings, which is where all the nitty-gritty but important details about your phone live. Other things you'll find lurking here include: software version (e.g. iOS 5), model and serial numbers, Wi-Fi address and network information.

Above: Siri settings include language and voice feedback options.

Software Update

Head here to find out if there's a software update for your phone. If there is you can update; if not, good for you: you're already up to date.

Siri

This is where you get to tell Siri how to behave. Choose the language Siri speaks, decide if you always want voice feedback, teach it more about you by providing personal information and choose if you want Siri to start working when you raise the phone to speak, bearing in mind that switching off Raise to Speak can save battery.

Data Roaming

If you're heading abroad and you haven't got a data plan with built-in allowances, then using your phone for data-heavy activities, such as surfing the web, checking your Facebook app or sending emails, can prove costly. In order to switch it off go into Settings > General > Celluar Data Roaming and toggle to Off.

Share Your Phone's Internet Connection

The iPhone's internet connection can be shared with a computer, iPhone or iPad, provided that your network allows it. In order to enable this, go into Settings and select Personal Hotspot and toggle the button on or off as required.

iTunes Wi-Fi Sync

You can sync your iPhone to a PC or Mac without needing to physically connect the device with a USB or Lightning cable (if you have the iPhone 5). If Wi-Fi syncing has been set up in iTunes (see page 41 for setting up iTunes), a Sync Now button will appear. Hit this and your devices will do their thing.

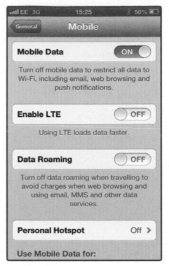

Above: Remember to switch off Data Roaming when travelling abroad to save costly charges outside of your price plan.

Customizing Your Keyboard

You can change the way your iPhone keyboard works; go into Settings > General > Keyboard and follow the instructions below to tailor your keypad settings.

Hot Tip

Before syncing your phone with iTunes over Wi-Fi, ensure that you have a full battery or that your phone is plugged into a power source so that there is enough juice to complete the sync. A half-synced iPhone is an unhappy iPhone.

- **Auto-Capitalization**: This capitalizes the first letter of every sentence after a full stop, question mark or exclamation mark.

- **Enable caps lock**: If caps lock is enabled, double-tapping the shift arrow key makes all letters you type appear in upper case LIKE THIS. Use wisely, though, as it is internet code for SHOUTING!

Above: Changing the keyboard settings allows you to customize your typing experience.

○ **Space bar double tap**: Turn this feature on and every time you double-tap the space bar, you'll enter a full stop and a space to end a sentence.

○ **International keyboard**: You can change to a foreign keyboard set up in Settings > General > International.

○ **Auto-Correction**: Switching this on makes the iPhone look for – and fix – typos and misspellings as you type.

Reset Settings

Only the brave, the desperate or the foolish mess around with the Reset settings – or maybe those about to sell their iPhone on eBay!

Luckily, you'll need a password before you can hit the big red buttons but here's what each of them does, should you decide to proceed.

○ **Reset all settings**: Resets your settings to default but doesn't affect your data or media.

○ **Erase All Content and Settings**: Deletes all your data from the phone and resets the settings to default.

- **Reset Network Settings**: Restores your phone's network settings to the factory defaults.

- **Reset Keyboard Dictionary**: Removes any words you may have added to the dictionary using the iPhone's intelligent keyboard.

- **Reset Home Screen Layout**: Returns your Home screen to the way it was when your phone left the factory.

- **Reset Location & Privacy**: Restores default settings for location and privacy.

Above: After entering a password you are able to begin resetting your phone settings.

Twitter

In Settings > General, hitting the Twitter button lets you add a new account. If you select Update, it will scan your contacts, adding Twitter handles and photos where available. You can also select which apps you want to use with Twitter.

Facebook

Within Facebook settings, you can turn on Calendar and Contacts settings. This will automatically add your Facebook pals to your phone Contacts, including profile pictures and email addresses, and it will also drop their birthdays into your iPhone Calendar. Please remember that the iPhone will try to match Facebook contacts with contacts already in your address book (for more details on managing Facebook contacts on your iPhone see page 111).

Hot Tip
Create your own keyboard short cuts to make typing faster. Go into Settings > Keyboard and Add Shortcuts. You'll be able to type things like GTBL and the full text 'Going to be late' will appear as a suggestion as you type.

TROUBLESHOOTING

Even good gadgets go wrong and the iPhone isn't immune to the odd glitch. If yours freezes or won't play nice, don't panic: there's generally a simple fix. This section will help you to overcome most of the problems you're likely to encounter.

CONNECTIVITY

The iPhone 4 and the iPhone 5 have come under fire for calling and connectivity problems. Bugs are often fixed by iOS updates but if you can't wait for new software, there are techy solutions on the web – be warned, though: these tend to be complicated to implement. So before you start meddling with DNS settings (DNS what? Exactly), here are some simple cures to most connection quibbles.

Wireless

Many iPhone 5 and iOS 6 users have reported problems concerning inconsistent Wi-Fi connectivity and poor speeds. This suggests wider issues with the iPhone 5's compatibility with some routers but do the following basic checks before you head back to the Apple store.

- **Is Wi-Fi enabled?** Tap Settings > Wi-Fi and make sure that Wi-Fi is turned on.

- **Connected to a network?** No? Pick an available Wi-Fi network from the list you'll find under Choose a Network and tap the one you want to join.

Left: If you are having connectivity problems, check your Wi-Fi settings.

- **Entered a password?** Some networks require passwords, indicated by a padlock icon. Double-check that your password is correct. Tip: If the Join button isn't selectable, the password is too short for that network.

- **Check the signal**: The Wi-Fi icon in the Status Bar shows a varying number of bars to indicate the signal strength. More bars equal stronger signal.

- **Weak signal?** Move closer to your router or Wi-Fi point and remember that brick walls and other interferences can affect the signal.

- **Can you browse the web?** Launch a browser to test the web connection. Navigate to an easy-to-load page like google.co.uk.

> **Hot Tip**
>
> **On public networks (hotels, airports, etc.), your browser will often redirect to a login page. Until you've signed in, the status bar will show you're connected but you won't be able to use the Wi-Fi network.**

- **Check your Wi-Fi network**: If the status bar shows you're connected but web content won't load, it suggests a problem with the Wi-Fi network you're using. If you're at home, check the cable connection to your Wi-Fi router or try connecting another device to the network to test it.

- **Reset network settings**: This will clear your 3G and Wi-Fi network settings, including saved networks, Wi-Fi passwords and VPN settings, so it is a last resort. Tap Settings > General and then scroll down and press Reset > Reset Network Settings. When the phone reboots, find and join the Wi-Fi network again.

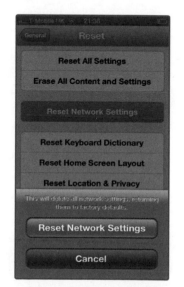

Right: Resetting network settings should be a last resort when solving connectivity problems.

Bluetooth

Having trouble pairing and sharing? Here are five simple fixes to get your Bluetooth working.

1. **Is Bluetooth enabled?** Go to Settings > General > Bluetooth and toggle Bluetooth On.

2. **Is your device paired?** In Bluetooth settings you'll see a list of items you've previously paired with your iPhone. Ensure the device you're trying to use is on this list.

3. **Is your device connected?** On the same list, ensure the word 'Connected' is displayed next to the device name. This shows you're communicating with it.

4. **Bluetooth headset controls not working?** In some cases the full range of features on third party Bluetooth accessories, such as the headset's volume controls, might not work. Sadly, not all products are fully compatible.

Above: The Bluetooth settings screen indicates which devices you are connected to.

5. **Switch Airplane Mode On and Off**: Go to Settings and toggle Airplane Mode On and Off; this will disable and restart your Bluetooth.

Phone

If you're having trouble making or receiving calls and texts, try the following.

1. **Make sure Airplane Mode is off**: Go to Settings and switch Airplane Mode Off.

2. **Check the signal**: In the top left-hand corner you'll see the signal bar. The more bars, the stronger the signal. If you're only seeing one or two bars, calls and texts might not work.

3. **Change location**: If you're indoors, head outside. If you're already outside, walking a few feet can make all the difference – chase that signal!

4. **Using an iPhone 4?** You may need to purchase a case to combat well-documented antenna issues. These prevented some people from making calls while holding the phone a certain way.

5. **Switch Airplane Mode On and Off**: Go to Settings and toggle Airplane Mode On and Off. This resets your wireless data connections and can flush out related problems.

6. **Restart your phone**: Yes, we know it's an old trick but it often works.

7. **Reset your network settings**: This clears your 3G and Wi-Fi network settings, including saved networks, Wi-Fi passwords and VPN settings, so it is a last resort. Tap Settings > General, scroll down and then press Reset > Reset Network Settings.

8. **Check your SIM Card:** With your phone switched off, use the SIM eject tool or a small paperclip to remove the SIM tray and take out the SIM Card and reposition. Reinsert the tray and restart the phone.

9. **Restore your phone:** Still no joy? The next step is to restore your phone (*see* page 24 to find out how this is done).

Above: If your status bar says 'searching' rather than displaying a series of bars, there may be a problem with your signal.

SYNCING

This section looks at problems that can occur while syncing your iPhone with iTunes, iCloud and any gremlins preventing your computer from recognizing the existence of your handset.

iTunes

When connecting your device to your computer – either via USB or Lightning connector for iOS devices purchased after September 2012 – an iPhone icon should appear in iTunes under Devices in the left-hand column. If this doesn't happen then here is what you can do.

Sleep/Wake button

- **Check your USB connection**: Always ensure your USB/Lightning cable is plugged into one of the main USB ports on your computer rather than a port on a keyboard or USB hub. If one port isn't working, try another. Also, check that the cables and USB ports aren't faulty.

- **Update iTunes**: Visit the Apple website at apple.com/itunes/download to make sure you have the latest version of iTunes. If not, follow the steps to update the software.

- **Restart your phone**: Press and hold the Sleep/Wake button and then slide the red slider to switch off your phone. Press and hold the Sleep/Wake button to restart it.

- **Recharge your phone**: Use the mains charger and bear in mind that if your phone was completely dead then the charging symbol may take time to appear.

Above: Restart the phone by pressing and holding the sleep/wake button, as indicated

○ **Restart your computer**: Switching your PC or Mac off and on again can sometimes fix the problem. Disconnect your phone before you do this.

○ **Uninstall and re-install iTunes**: It's long-winded but re-installing iTunes software can help. This process is managed differently, depending on your version of Windows (visit apple.com/support for more details).

Unable to Sync Photos in iTunes

If you're using iTunes rather than iCloud, you may discover problems syncing photos. For example, the progress bar might freeze or you might be told there is not enough space. If that happens, try the following.

Below: iTunes photo pane showing photo syncing options.

1. **With iTunes open:** Select the device, then select the Photos pane and turn off Photo syncing.

2. **Sync your device:** Remember that this will remove all synced photos from your iPhone. Any you've taken since the last sync may be lost.

3. **Disconnect your device:** Then update your iTunes.

4. **Delete your iPhone photo cache:** This folder stores iPhone-optimized copies of your pictures. Use your computer's Search Files function to find it. Deleting won't affect the originals and a new copy will be created next time you sync.

5. **Connect your device to iTunes:** Select the device, then the Photos pane and turn Photo syncing back on. After that you can re-sync your device.

> ## Hot Tip
> Using iCloud rather than iTunes to sync and back up photos means that your pictures will be automatically updated when you join a Wi-Fi network (*see* page 42 for more iCloud benefits).

iTunes Syncing Over Wi-Fi

Since the launch of iTunes 10.5, you can sync your iPhone with iTunes over the air and go fully computer-free. For this to work you'll need a device running iOS 5 or later and be connected to the same Wi-Fi network as your computer, which needs to be running iTunes 10.5 or later. If that's all present and correct and you're still having problems, there are some crucial checks you need to make.

○ **Ensure Wi-Fi sync is enabled:** In order to sync your content to your iOS device via Wi-Fi, you need to ensure that this option has been enabled in iTunes (*see* page 41 for more details on syncing and iTunes).

○ **Quit iTunes and restart**: Close down your iTunes software and reopen it, then restart your iPhone.

○ **Restart your network router**: This will vary from device to device but the easiest way is generally to remove the power cable and then reinsert it after 10 seconds. During this time your network will be down.

○ **Check your network connection**: Ensure the problem isn't with the Wi-Fi network by connecting another device to the internet on the same network (*see* page 226 for more tips on solving connectivity and Wi-Fi problems).

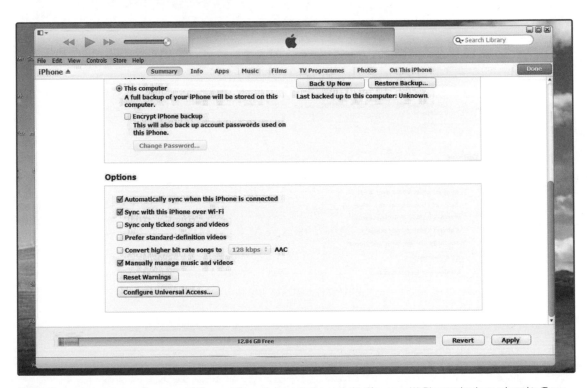

Above: If you want to sync your iPhone over Wi-Fi you need to ensure that the 'Sync with this iPhone over Wi-Fi' option has been selected in iTunes.

○ **Check your firewall settings:** If your computer has a firewall running, this can prevent it from communicating with your iPhone. In order to change your firewall settings, you'll need to refer to your manufacturer or anti-virus software provider for assistance.

Not Enough Free Space

Whether you're syncing over Wi-Fi or via USB/Lightning, you might encounter the following message: 'iPhone cannot be synced because there is not enough free space to hold all of the items in the iTunes Library (additional space required)'. In order to fix this, try turning off the automatic syncing function in iTunes, as shown here.

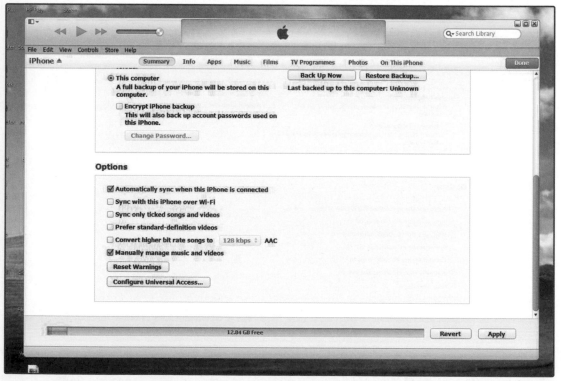

Above: If you run out of free space you should deselect the option to Automatically sync when this iPhone is connected. Instead you should select the option to Sync only ticked songs and videos.

- **Select your iPhone**: Look under Devices on the left-hand side of iTunes, find your iPhone and click the Summary tab.

- **Turn off Auto-Sync**: Deselect Automatically sync when this iPhone is connected and select the Sync only ticked songs and videos checkbox.

Hot Tip

Having trouble syncing over Wi-Fi? Check that your iPhone is connected to the same Wi-Fi network as your Mac or PC, as it won't work if it isn't. Your iPhone will also need to be plugged into a power source.

- **Click Apply**: This will sync the changes to your iPhone.

- **Sync less data**: If syncing your entire music library exceeds the memory capacity of your iPhone then choose Selected playlists to transfer rather than All songs and playlists under the Music tab in iTunes. You can also manage apps, films and iBooks in the same way.

iCloud

iCloud is a great tool for backing up all of your favourite content and being able to access it from any Apple device but it's also a potentially complicated beast if it goes wrong. There are far too many troubleshooting issues to cover here in detail (you can get detailed help for most issues at apple.com/support/icloud) so, for now, we'll show you how to spot and fix the most common troubles.

Can't Sign into iCloud?

Make sure you're using the Apple ID email address you used when you set up your iCloud account. Also remember that passwords are case sensitive so check you don't have caps lock on. If you've forgotten your password you can reset it online at appleid.apple.com.

Automatic Backups Won't Start?

Automatic backups only start when your iPhone is plugged into a power source, connected to Wi-Fi and the screen is locked. If that still doesn't work, you can start a backup manually by tapping Back Up Now in Settings > iCloud > Storage & Backup, but bear in mind that this also requires a Wi-Fi connection and power.

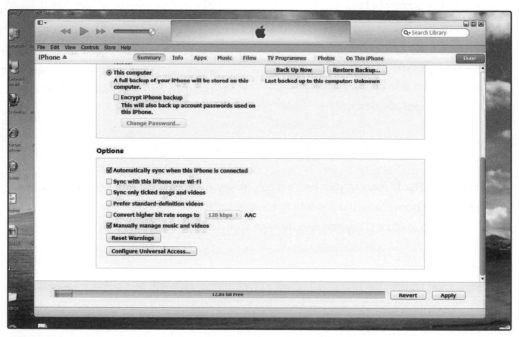

Above: You can check and adjust your back up settings through the Summary section of your iTunes account.

Hot Tip

Ensure you have at least 50 MB free space available on your iPhone before you attempt to back up. If you have no space available, iCloud Backup can fail. Verify how much free space you have available on your iOS device in Settings > General > About.

Out of Cloud Storage Space

If you're told you don't have enough cloud storage space to back up all of your content to iCloud, try excluding your Camera Roll and other large data items from backups. Go to iCloud > Storage & Backup > Manage Storage and choose your device. Alternatively, you can remove larger files, such as videos you've watched, from your device to make more space. If that's not fixing the problem, you can always buy more storage to add to the free 5 GB provided by Apple.

My Photos Are Not Appearing in My Photo Stream

From your iPhone's Home screen, select Settings > iCloud > Photo Stream and make sure that the slider is On. Then make sure that Wi-Fi is switched on and that you're connected to a wireless network. Photo Stream won't upload photos from an iOS device until the Camera app is closed on the device you used to take the photo, so check this. Also ensure your iPhone battery hasn't dropped below 20 per cent, as Photo Stream downloading and uploading are disabled when the battery reaches this threshold.

Above: You can check how much storage you still have available.

Above: You can check and adjust your photo stream settings.

APPS

Most of the apps in the App Store work well – most of the time. However, even the best maintained apps can be prone to bugs; from crashing to sluggish performance, here's what to do when your apps aren't working.

Close the App

If an app gets stuck during load, you can bump-start it. Double-tap your phone's Home button to see a list of everything running in multitasking. Scroll right to find the rogue app, then press and hold until a badge with a '–' appears in the left-hand corner of the icon. Tap the badge and the app will close fully. Then reopen the app as normal.

Reset Your iPhone

Resetting fixes 95 per cent of glitches. Hold down the Home and Sleep/Wake buttons simultaneously for 10 seconds. Keep holding the buttons until the screen goes blank and the Apple logo appears. Let the buttons go and the phone will restart. Then open the app again.

Above: Multitasking screen showing which apps are live.

Delete and Re-install the App

If the first two steps failed, you might have to delete the app. Unfortunately, you'll lose that app's local data such as games' high scores. If that's OK then press down on the app icon until an 'x' appears over it; press it and confirm you want to delete the app. Once it's gone, you can then reinstall it.

BATTERY

The battery life of smartphones has always been a bugbear. You could blame the manufacturers but many of the problems we encounter with battery life are actually due to how much and how often we use our handsets. We expect to be able to use them for longer, doing ever more complicated tasks, often with two or three apps running at the same time. In tests the iPhone 5 has been shown to last more than seven hours so if you're getting less these tips and tricks might help your phone to stay alive for longer.

Loses Power Quickly

If your iPhone battery is winding up empty quicker than you would like, there's a chance you've got a faulty battery. However, it's more likely that your settings and usage are drinking more power than necessary. Here are some simple ways to make your iPhone last longer.

- **Drain the battery**: About once a month, or if you're having problems, it's recommended that you completely drain your iPhone's battery. Let it run until it shuts down on its own and then charge it back up to full.

- **Close background apps**: Apps carry on running in the background unless you close them. Double-tap your phone's Home button to see a list of everything running in multitasking.

Scroll right to find the rogue app, then press and hold until a badge with a '–' appears in the left-hand corner of the icon. Tap the badge and the app will close fully.

- **Cut the number of Push Notifications**: Push notifications are notifications sent by apps straight to your phone. If lots of them are enabled, they can wear down your battery. Go to Settings > Notifications and click on the apps from which you don't need notifications.

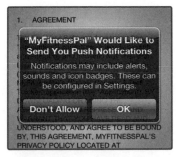

- **Using your iPhone in an area with weak 3G signal**: Your phone works harder when it's struggling to find a 3G signal. If you're going to be in an area with bad coverage for a long period, switch on Airplane Mode.

Above: Screenshot showing the options given when you first download an app.

- **Adjust screen brightness**: Switch on the Auto-Brightness function in Settings > Brightness & Wallpaper and the iPhone will adjust itself depending on light conditions.

- **Switch on Airplane Mode if you don't need to be connected**: Airplane Mode switches off your 3G, Wi-Fi and Bluetooth connectivity, thus cutting out power-draining updates.

- **Cut the number of apps that can use Location Services**: A lot of apps like to know where you are but if you don't need Twitter to stamp your Tweets with a location then switching this off can save power. Go to Settings > Location Services and choose for which apps you want to use location.

- **Turn off Raise to Speak function**: Many iPhone users have found that having the Raise to Speak function switched on causes battery drain. Turn this off by going to Settings > General > Siri and turn off Raise to Speak.

- **Switch email to Fetch, not Push, and set a longer gap between updates**: If your email account is set to Push, it updates when a new email comes in, using a little more power each time, whereas Fetch lets you check for emails periodically. Set a longer gap between checks and you'll save battery.

Above: To save battery turn off Push and select Fetch in your mail settings.

○ **Turn Wi-Fi off if you're not using it**: The constant search for Wi-Fi networks will drain your power. As with 3G, constant scanning for a Wi-Fi network drinks juice.

○ **Update your iOS software**: Each version of iOS tends to deliver battery life improvements. Making sure you're running the latest version means you can take advantage of any improvements Apple has released.

○ **Cut the number of Push Notifications**: These are notifications sent by apps straight to your phone and if lots of them are enabled, they can wear down your battery. Go to Settings > Notifications and click on the apps from which you don't need notifications.

FROZEN iPHONE

iPhones freeze for a number of reasons. Apps are often the cause and closing a rogue app can get you going again. However, if your phone completely locks up, follow these simple steps to get it back up and running.

Recharge

Make sure your phone is fully charged. If you can, turn it off while it's charging and use the mains charger to speed things up. If a full battery doesn't cure the seizure, it's time to restart.

Restart

Press and hold the Sleep/Wake button and then slide the red slider to switch off your phone. Then press and hold the Sleep/Wake button to fire it back up. If this hasn't fixed the problem, it's time to reset.

Reset

Resetting your iPhone is the equivalent of rebooting a crashed computer. It's a simple technique that shouldn't affect your data but can resolve a host of issues. Simply press and hold the Sleep/Wake button while simultaneously pressing and holding the Home button. You should see an Apple logo appear. After that you can let go and your phone should restart.

RESTORE

Restoring your phone is not something you do lightly but if you've tried recharging, restarting, removing content and resetting your settings and all that has failed, then it might be your only option. Here are step-by-step instructions to follow but, before you start, you should back up your phone if you can.

Hot Tip

Restoring your iPhone erases all data and media, and resets your settings to factory defaults. However, provided the problem you're fixing has been dealt with, you should be able to reload everything from your most recent backup next time you sync.

1. Connect your phone to your computer like you would if you were syncing.
2. When iTunes opens, select the Summary tab and click the Restore button.
3. Re-sync your phone to restore your data.

RECOVERY

If recharging, restarting, resetting and restoring haven't fixed your fault, the final option is to put your phone into Recovery Mode – here are the step-by-step instructions to follow.

1. If you have a Lightning or USB cable connected to your iPhone, disconnect it but leave the other end plugged into your computer.
2. Turn off your iPhone by pressing the Sleep/Wake button and sliding the red slider.
3. Reconnect the Lightning or USB cable to your iPhone while simultaneously pressing and holding the Home button and your iPhone should come on.

4. Keep holding the Home button and only release it when the Connect to iTunes screen appears.

5. iTunes should launch automatically but if it hasn't, open it manually.

6. Restore your phone following the steps in the Restore section (*see* page 241).

RUNNING OUT OF SPACE

Once you start downloading apps, games, music, films and photos, it's easy to fill up your iPhone's storage – even if you're the proud owner of a 64 GB model. The good news is that there are many ways to manage your content more smartly to free up space.

Managing Storage Space

○ **Find out what's using your storage**: Go to Settings > General > Usage. You'll find details of how much used or free space you have, plus a list of content and details of how much capacity each item is taking up. Use this to identify large apps you no longer use.

Above: Check the usage screen in your settings to see what's using your storage.

○ **Delete unused apps:** Remove games you've completed or apps you no longer use by pressing and holding their icon on your Home screen until it wobbles, then tap the 'X' and confirm. Please remember that apps you've purchased can be reloaded as long as you've not deleted them from your iTunes Library.

○ **Remove Old Videos:** Video content uses a lot of space. Delete old videos that you've watched and no longer need on your iPhone.

○ **Standard Definition vs High Definition:** Choosing to download standard definition videos over their larger hi-def counterparts saves space. Set this up via Options in iTunes by checking the box next to 'Prefer standard definition videos' and SD content will be preferred over HD.

ACCESSORIZING

The Apple ecosystem is a marvel of modern business. Thousands of small enterprises and innovators work tirelessly to provide everything, from cases and docks to headphones and bike mounts, to help make your iPhone more useful than ever – and of course to earn a few pounds. This section covers what you might need.

COVERS

The iPhone is fragile; drop it and that stunning screen is liable to crack. Fortunately, protecting it is cheap and gives you another chance to personalize your phone. There are many different styles of case on the market, from bumpers that cover just the outer edges of the handset to those that clip over the back or even options that cover – and waterproof – the entire phone. Apple-endorsed cases can be found in Apple Stores but a simple Google search also delivers a world of choice.

CONNECTION KIT

Connectivity and cables have become an issue for Apple. The iPhone 5 saw the introduction of the Lightning interface: a new system of cables for charging and connecting your iPhone to laptops, TVs and docks. Older models of the iPhone, iPods and iPads use the 30-pin cable that has become synonymous with Apple products. This means that there are an enormous number of docks, chargers and accessories that aren't compatible with the iPhone, unless you buy a 30-pin-Lightning adapter.

Above: Use a cover to protect your precious iPhone; there are many kinds available.

Above: Non-Apple headphones may be better quality.

SOUND

As your iPhone is also an iPod and often a hub for music, video and entertainment, finding the right audio peripherals to make the most of your media is important.

Headphones and Headsets

The headphones that come boxed with the handset pre-iPhone 5 aren't very good. Apple has just improved this for the iPhone 5 but it's advisable to invest in a pair of non-Apple headphones for better audio quality. There are many options and you can choose between wired or Bluetooth.

AirPlay, Bluetooth and Docks

Apple is pushing its new wireless streaming format, AirPlay, and with the introduction of the Lightning connector for the iPhone 5, millions of docking speakers have been made virtually redundant. Both Bluetooth and AirPlay let you send music to a compatible speaker system over the air. Most reputable audio manufacturers now offer a range of systems to make this easy to do.

Hot Tip

If you want to guarantee that your headset of AirPlay speakers will work perfectly with your iPhone, look out for Apple's endorsement messages on the packaging. Without these, some functions might not work.

STYLUS

If you want a little more pen-style control, you can purchase a stylus to type, tap and scroll on your iPhone. Apple offers two types: the Soft-Touch Stylus and the Pogo Stylus, but there is a wider selection available.

Above: You can use a stylus pen to scroll, type and tap on your iPhone screen.

MEMORY

In order to have your iPhone performing at its best, you need both RAM (random access memory) and free storage space. The way the iPhone is built means that you have fairly limited options for maximizing this, but here are a few things you can do.

IMPROVING MEMORY

Closing apps can help to boost memory. You may not be using an application, but if you don't manually close it, it still runs in the background, drinking power, leeching memory and potentially using valuable mobile data. In order to close the application, double-click the Home button to bring up the Multitasking Bar and then hold your finger down on one of the app icons until a red circle appears next to it. Click this circle to close the app.

INCREASING MEMORY

Unlike some smartphones, it's not possible to extend the iPhone's storage space with an external memory card. The only way to increase available space is to delete the apps you never use and remove old content you no longer need, such as videos and podcasts. You can manage this on the handset or via iTunes but be aware: anything removed locally on the device will reappear unless you also remove it from your iTunes (see the section about managing storage space on page 234 to learn how).

Left: Increase memory by deleting little used apps and removing old content that you no longer need but is indicated as taking up a lot of room in your Usage.

SECURITY

The iPhone is an expensive piece of technology. Keeping the physical product safe is essential but it's just as important to protect the data and information stored on your phone. From names and addresses to documents and your access to web-based accounts, your phone is a key to your life and security is paramount.

Above: You can choose your own passcode to increase security.

PASSWORD PROTECTION

Like most things techy, the iPhone comes with a password so you can protect yourself and your information. Although it's another number or magic word to remember, it's advisable to assign a passcode. It's easy to do and, should your phone fall into the wrong hands, you'll be thankful you did.

Passcode Lock

You can set a password to prevent unscrupulous people from unlocking your phone and sending rude texts to your mum/boss/teacher. Go into Settings > General > Passcode Lock and enter a four-digit code. You'll be asked to verify the code to ensure it matches.

Above: Having set an initial passcode you can change it at any time.

Increased Security Passcode

If you want to set a more complex passcode, you can switch off Simple Passcode. You'll then get the option of assigning a longer password that combines numbers, letters and special characters.

Above: You will be asked to enter your passcode when you unlock your phone.

Failed Passcode Data Dump

You can set the iPhone to automatically erase all your data if someone makes 10 failed attempts at unlocking your phone with the passcode. Be warned, though: all your media, data, settings and information will be deleted so think carefully before you activate this.

Auto-Lock

Want to make your phone lock itself if it has been sitting idle? Go into Settings > General > Auto-Lock and you can set the amount of time that elapses since your phone was last used before it automatically locks the display. Your choices range from five minutes down to one minute.

Above: You can select the duration of time after which your screen will automatically lock.

SIM Locking

Any data that might be stored on your SIM Card – anything from phone number to photos – can also be protected with a PIN number. Turn on SIM PIN and enter a password; this will prevent anyone else from using it in another phone without knowing the magic code.

Restrictions

The iPhone's restrictions tools let you dictate what can and can't be done with your phone. From locking the Safari web browser to preventing new apps being loaded on to the device, it's possible to manage the phone in a way that adds layers of safety, security and parental guidance.

Hot Tip

Find your lost iPhone. Go to Settings > iCloud and turn on Find My iPhone. Then if you lose your handset, log into your iCloud account via the web and you should see on a map where it is – fingers crossed!

ENCRYPTING BACKUP

Whenever you connect your iPhone to iTunes to sync, update or restore your device, most of your essential data is backed up to your computer's hard drive or to iCloud. This includes photos, text messages, notes, contact favourites and some settings. It's possible to keep this data safe from unwanted prying eyes by encrypting your backups for an added layer of security.

Encrypt Your Data

If you choose to back up your iPhone to your computer you get the option to encrypt and password-protect your data. It's simple to do. Just connect your handset to your computer and open iTunes. Select your iPhone in the sidebar on the left of iTunes and click the Summary tab. Click Back up to This Computer and check the box that says Encrypt iPhone Back Up.

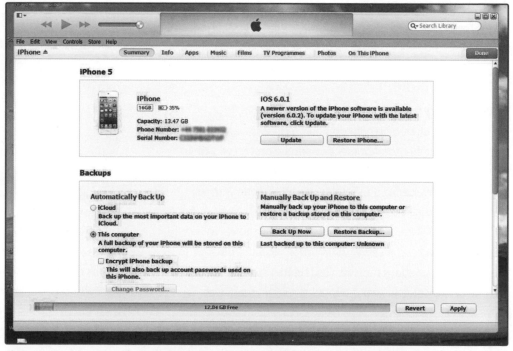

Above: You can choose to encrypt your iPhone data by selecting Encrypt iPhone backup option from the menu in the iTunes summary tab; this increases your iPhone security.

VPN

A VPN, or Virtual Private Network, is most commonly used to provide secure access to a company's network behind a firewall. A VPN creates an encrypted internet connection that acts as a safe channel for sending and receiving data. You're most likely to have encountered this when sending emails on your work account from a phone outside of the office.

Setting up a VPN on Your iPhone

Setting up a Virtual Private Network is one of those things that sounds more complicated than it is. However, if you can get it, we'd advise seeking help from your company's IT experts. If you can't get help, at the very least you'll need them to provide some key information before you start, including which protocol to use and the main configuration settings. If you're flying solo, once you're in possession of those vitals, here's what to do.

Above: You can set up a VPN on your iPhone.

○ **Hit settings and go to Network**: Select VPN and then Add VPN Configuration.

○ **Choose the protocol**: You'll be asked to select which VPN protocol you wish to use. Choose from: L2TP, PPTP or IPSec VPN.

○ **Fill in the form**: Much of the setup is filling boxes. When prompted, enter the server information, account details, password and encryption level (if one has been provided) – plus any other details you've been given by your IT people.

○ **Switching a VPN on or off**: Once the initial setup is complete, you can toggle VPN on or off as you wish via Settings.

Hot Tip

You can use a VPN to watch the iPlayer abroad or even to access Facebook in China. Running your internet traffic through a different network makes your iPhone look like it's in the UK, while you're really typing away in a hotel suite in NYC.

TOP 100 APPS

This is our list of the top 100 apps we think that every iPhone user should have, or aspire to have. Some are essentials; others are those that you might use only occasionally but are what brings your iPhone to life. Have a look through our picks and see what your iPhone is missing out on.

Shopping

1. **eBay:** Sell or buy and find out instantly when a bid has been successful. **Free**

2: **Amazon Mobile:** Use the iPhone camera to take a picture of a product to see if it's in stock. **Free**

3. **Tesco:** The supermarket chain's app lets you do the shopping on the move. **Free**

4. **Ocado:** Use a barcode scanner to add items to your shopping list. **Free**

5. **Groupon:** Get special offers on experiences and events, depending on your current location. **Free**

6. **Gift Plan:** Reminds you when a special event is looming, and gives you gift ideas. **£1.99**

Connectivity

7. **Safari:** Apple's native web browser. **Free**

8. **Opera Mini:** Offers fast page loading and jumps quickly from multiple webpages. **Free**

9. **Chrome:** Supports unlimited tabs and can send pages from your computer to your iPhone. **Free**

10. **Gmail:** Google's email client will let you read email threads as conversations. **Free**

11. **Skype:** Make free voice calls, video conversations and instant messaging. **Free**

12. **FaceTime:** Apple's video calling app uses the iPhone's front-facing camera. **Free**

Watch and Listen

13. **BBC News:** Access content from one of the world's most reputable news services. **Free**

14. **Sky Go:** Sky subscribers can watch both live and on demand content. **Free**

15. **BBC iPlayer:** Catch up on BBC TV programmes and radio from the past seven days. **Free**

16. **4oD:** Channel 4's TV on demand service lets you catch the past 30 days of programming. **Free**

17. **Sky+:** Change your iPhone into a touchscreen remote control. **Free**

18. **Virgin TV Anywhere:** Virgin TiVo customers can watch live TV over a Wi-Fi connection. **Free**

19. **TVCatchup:** Stream over 50 Freeview channels with a Now and Next view. **Free**

20. **Netflix:** Pay a monthly subscription and watch TV shows and films in full HD. **Free**

21. **LOVEFiLM:** Choose from over 70,000 TV shows and movies from the US and the UK. **Free**

22. **Apple Remote:** Take control of iTunes or Apple TV and select playlists or adjust volume. **Free**

23. **YouTube:** Browse and watch millions of videos with new voice search support. **Free**

24. **Google Drive:** Store and upload music and video that can be accessed anywhere. **Free**

25. **TED Mobile**: Access global conferences where experts talk about ideas in 20 minutes or less. **Free**

26. **Spotify**: Premium subscribers stream music over 3G and Wi-Fi, and save playlists off-line without being connected to the internet. **Free**

27. **TuneIn Radio**: With access to over 60,000 global radio stations, you can pause live streams to re-listen to favourite shows. **Free**

28. **iPlayer Radio**: The iPlayer for audio, it has access to 300 UK radio stations including your favourite BBC stations. **Free**

29. **Shazam**: Hold the Shazam app up to a song and it will tell you all you need to know about the track. **Free**

Reading

30. **The Guardian**: Offering access to the latest content from the British newspaper. **Free**

31. **Newsstand**: Subscribe to digital newspapers and magazines, and the latest issues will be added to the virtual shelf. **Free**

32. **Pulse News**: Bring your favourite news websites into one place; save stories and read them off-line. **Free**

33. **Zinio**: A catalogue of magazines to read on or off-line, single issues or subscriptions can be bought through your iTunes account. **Free**

34. **Instapaper**: Collect news stories from RSS feeds and the web, they will be stripped down so they are easier to read. **Free**

35. **Pocket**: Pocket syncs to your device so you can catch up on articles when it is convenient. **Free**

36. **iBooks**: Apple's official ebook store lets you buy and read a variety of books. **Free**

37. **Kindle**: Read Kindle books, newspapers and magazines on your iPhone. **Free**

Social Media

38. **Facebook**: The friend-collecting site will let you update your status, check your news feed and receive notifications on posts. **Free**

39. **Twitter**: Tweet on the go with the new 'Discover' feature helping find suitable tweets and followers. **Free**

40. **LinkedIn**: Giving you access to your entire professional network. **Free**

41. **Zeebox**: Combining social networking with live TV, view additional content about programmes and share comments. **Free**

42. **Tweetcaster**: An alternative to the official Twitter app, Tweetcaster lets you post simultaneously to Facebook and Twitter. **Free**

43. **eScrap App**: Build a scrapbook by uploading photos on to 50 different layouts and share your best on Facebook and Twitter. **Free**

44. **Flipboard – Your Social Magazine**: Turn customizable news stories and social networking updates into a glossy magazine. **Free**

Photos & Video

45. **Hipstamatic**: Produce retro-looking Polaroid style photos with your iPhone. **£0.69**

46. **Instagram**: Add filters to photographs to give them a vintage look, and share with other users via Facebook or Twitter. **Free**

47. **Quad Camera – Multishot**: Have fun with photos by taking up to eight pictures simultaneously, add effects before sharing. **£1.49**

48. **Adobe Photoshop Touch**: Includes many of the creative tools in the desktop version to modify and enhance images. **£6.99**

49. **Dropbox**: Access files wherever you are, as well as off-line, upload video and photos in bulk. **Free**

50. **Pinterest**: Create boards and 'pin' images from the web to show the world what inspires you. **Free**

51. **Flickr**: The photo sharing app that now supports video will let you upload multiple content. **Free**

52. **Moonpig**: Upload photos from your iPhone to create a unique card that can then be sent to the desired address. **Free**

Geo-Location & Travel

53. **TomTom UK and Ireland**: Turn your iPhone into a fully-fledged sat nav with turn-by-turn voice instructions. **£39.99**

54. **AA Safety Camera**: Avoid picking up fines with this warning system that will alert you to safety cameras. **£0.69**

55. **Find my Friends**: Locate contacts with an iPhone, iPad or other Apple device to make meeting up easier. **Free**

56. **Google Earth**: Search places from around the world with detailed maps. **Free**

57. **The Night Sky**: Using the iPhone compass and GPS, hold the app up towards the sky to identify stars and constellations. **£0.69**

58. **Pin Drop**: Never forget a memorable place by dropping a pin to bookmark locations. **Free**

59. **Trip Advisor**: Offers a complete off-line guide to cities, with visitor attraction details. **Free**

60. **Flight Track**: You can check the status of your flight: real-time departures, gate closing times and alternative flights. **£2.99**

61. **Skyscanner**: Find affordable flights covering over 1,000 airlines and purchase the ticket directly through the app. **Free**

62. **British Airways**: Flying with BA? You can check flight information and use your iPhone as a boarding pass. **Free**

63. **the trainline**: Check live train times, view your next train home and buy tickets. **Free**

64. **Vue Cinemas**: This simple movie-booking app lets cinema goers locate the nearest cinema via GPS. **Free**

65. **Odeon**: This app will help you find your nearest Odeon cinema and the films being shown. **Free**

Sport & Fitness

66. **ESPN Goals**: The first official place where you can catch up on all the Premier League goals. **Free**

67. **Fanatix**: A personalized feed of real-time sporting news which can be tailored to your favourite teams. **Free**

68: **British Military Fitness**: Offers a host of instructive videos and audio shape intense workouts that will help you get in shape. **£2.99**

69: **miCoach**: Prescribing training plans for different sports, GPS-based pace zones will help to push you further. **Free**

70. **Instant Heart Rate**: Place your index finger on the iPhone camera to read your pulse. **£0.69**

71. **Nike+Running**: Track and map your running activity using GPS and the iPhone accelerometer, and stay motivated with power songs. **Free**

Games

72. **Shaun the Sheep: Fleece Lightning**: Your challenge is to outrun 80 other farmyard animals. **£0.69**

73. **Temple Run**: The reaction-style adventure game tests your ability to avoid obstacles at speed as you are chased by a gang of angry apes. **Free**

74. **Cut the Rope**: Get the candy into the Om Nom monster by cutting ropes in this addictive puzzler. **Free**

75. **Whale Trail**: The psychedelic game puts you in control of a flying cartoon whale that has to consume bubbles and stars in the sky. **Free**

76. **Angry Birds**: Take out all of the pigs by flinging birds with different abilities from a slingshot in this massive mobile hit. **Free**

77. **Atari's Greatest Hits:** Retro gaming fans can play 100 Atari classics including Asteroids, Pond and Centipede. **Free**

78. **Draw Something:** The social game, which is like Pictionary, involves creating a picture based on a suggested word that friends must try to guess correctly. **Free**

Food & Restaurants

79. **Foursquare:** Rewarding you for exploring your surroundings, check in to places and earn badges to get exclusive deals and discounts. **Free**

80. **Jamie Oliver Food Guide:** The Naked Chef picks 1,000 of the best places to eat and shop for great food. **£2.99**

81. **Nigella's Quick Collection:** The gastronomic guide includes 70 recipes covering the best comfort foods and video tips of difficult meals. **£4.99**

82. **Hummingbird Bakery Cake Days:** Recipes and step-by-step instructions as well as a feature that allows you to clap to the next step for when your hands are full. **£2.99**

83. **Wine Navigator:** Learn how to pair the perfect wine with a meal and the ideal way to serve it. **Free**

84. **Kitchen Pad Timer:** Timers display oven and stove temperatures with an alert to remind you when to take something out. **£1.49**

To-Do List

85. **Evernote:** Take care of all note-taking from simple to-do lists to recording voice memos, all can be shared and accessed from evernote.com. **Free**

86. **PrinterShare Mobile:** Send documents and web pages to a nearby Wi-Fi-enabled printer to keep a hard copy for your records. **Free**

87. **Captio:** Taking the fuss out of reminding yourself to do something, Captio instantly sends notes, pictures and emails to your own inbox. **£1.49**

Reference

88. **HowStuffWorks:** A huge online archive of blogs, podcasts and videos to browse, and share facts on Facebook and Twitter. **Free**

89. **Dictionary.com:** You can check references and spellings online or off-line with audio pronunciations. **Free**

90. **Google Translate:** Translate phrases, sentences and words into over 60 languages after you have either typed them or spoken into your phone. **Free**

91. **IMDb Movies and TV:** Comprehensive source for information about every film and TV show ever made. **Free**

92. **Wikipanion:** Wikipanion remembers where you last left the page and keeps a history of all the pages you've searched for previously. **Free**

93. **QI:** Pulling together a library of books that include facts from the TV quiz show. **£2.99**

94. **Swype Type:** Making texting a speedier process, swipe across the keyboard to make words with predictive dictionary support to aid accuracy. **£0.69**

Finances

95. **Coin Keeper:** Monitor your finances by seeing how much you spend, tracking expenses and setting weekly budgets. **£3.99**

96. **Keybox:** Store all of your passwords, log-ins and bank account details securely in one place. **£1.99**

Useful Stuff

97. **UK Postage Calculator:** Get the right stamps on packages and letters by calculating the price depending on weight and the destination. **£0.69**

98. **iHandy Level Free:** This app can turn the iPhone into a spirit level as well as other DIY tools, such as a pendulum and a ruler. **Free**

99. **GloveBox:** Keep track of important vehicle details, from MOT and Road Tax renewal times **£0.69**

100. **Fake-A-Call:** Trying to get out of a boring meeting or presentation? Set times for your iPhone to call you. **£0.69**

INDEX